SUPERSTORMS

EXTREME WEATHER IN THE HEART OF THE HEARTLAND

TERRY SWAILS

FARCOUNTRY
PRESS

Acknowledgements

Bob Dylan said that you don't need a weatherman to know which way the wind blows. While there's a prevailing truth to the statement, many of you find that it helps. As a result, I've been able to live my life with my head in the clouds, pursuing the storms that I love. To all of you who have put faith in my forecasts, this book is for you.

Many people have earned my gratitude by contributing in some way to the development of this book. First and foremost, my heart goes out to my wife, Carolyn, who single-handedly raised our child while providing the support and love necessary to complete this project. To Boo, Rose, and Luann, your presence over the years has made the hard times easier and the good times great. And to Eden Malone, for opening the door of parenthood and bringing me the joy and purpose that goes along with it. To you, I dedicate this book.

I would also be like to express special thanks to Steve Gottschalk, who performed hours of research to find the trinkets of gold this book is based on. Credit also goes to NOAA, the Storm Prediction Center, and the offices of the National Weather Service (especially Davenport and its staff). I would also like to express my gratitude to Iowa State Climatologist Harry Hillaker, meteorologist Bill Nichols, photographer Jim Reed, and the many others who contributed valuable information, insights, and pictures. A special thanks also to author John L. Stanford and Iowa State University Press for permission to use the Camanche tornado account from Stanford's 1977 book, *Tornado*.

My gratitude also goes out to the *Quad-City Times,* in particular Terry Wilson and Roy Booker. Last but not least, a special thanks to Kathy Springmeyer, Caroline Patterson, Shirley Machonis, and the staff of Farcountry Press for the opportunity to make this dream come true.

Temperatures throughout are given in degrees Fahrenheit.

ISBN-13: 9781560373322
ISBN-10: 1-56037-332-6

© 2005 Farcountry Press
Text © 2005 Terry Swails

Library of Congress Cataloging-in-Publication Data

Swails, Terry.
  Superstorms : extreme weather in the heart of the heartland / by Terry Swails.
    p. cm.
  ISBN-13: 978-1-56037-332-2
  1. Climatic extremes—Middle West. 2. Natural disasters—Middle West. 3. Weather. I. Title.
  QC981.8.C53S93 2005
  551.55—dc22

                    2005019076

Front cover: PHOTO BY DOUG RAFLIK
Back cover: PHOTO BY MIKE HOLLINGSHEAD

For more information on our books, write Farcountry Press, P.O. Box 5630, Helena, MT 59604; call (800) 821-3874; or visit www.farcountrypress.com.

Created, produced, and designed in the United States.
Printed in Korea.

10 09 08 07 06 05   1 2 3 4 5 6

Table of Contents

CHAPTER

1

*Sunshine is delicious, rain is refreshing, wind braces us up, snow is exhilarating;
there is really no such thing as bad weather, only different kinds of good weather.*
—JOHN RUSKIN

Introduction: Into the Looking Glass

Forty years ago, my mother sat me down on the couch to watch *The Wizard of Oz*. Zipped up tight in my pajamas, my sister at my side, I didn't realize that my life was about to change forever. As the throbbing black cloud slithered down from the sky, I stood up from the couch and watched, fists clenched, as Dorothy, Toto, and Dorothy's house went spinning across the TV screen.

When the wind subsided and Dorothy awoke, she found herself in another land. So did I. It had nothing to do with the movie's wizards or flying monkeys, good witches or bad. I was transported by the violence and beauty of that tornado that swept Dorothy out of Kansas and into Oz. In the blink of an eye, I had discovered my passion for weather—for blizzards, lightning, hailstones, and twisters. I didn't choose weather; weather chose me.

Nearly 40 years later, I find myself in my kingdom of radars, satellites, and computers, delivering the tornado warnings that might well have spared Dorothy her journey to Oz. In a twist of fate, I am now the voice of reason people depend on when the skies turn dark and stormy. My words have the ability to make the difference between life and death. When I present the specifics about a threatening

storm, I am ever-mindful that my message has to be focused and razor-sharp in its clarity. It's a moment of truth: at that instant, I have the most important job in the world. And that is just fine by me.

Before I was ten, I knew the names of all the local television forecasters and at what times and on which stations they appeared on. With remote control devices still years away, it was not unusual to see me frantically cranking the dial on our big television set to hear what each forecaster had to say.

My idol was Conrad Johnson, a weatherman who worked at WMT-TV in Cedar Rapids, Iowa. When Johnson spoke, I listened. I loved his confident, easy delivery and the accuracy and precision of his forecasts—qualities that

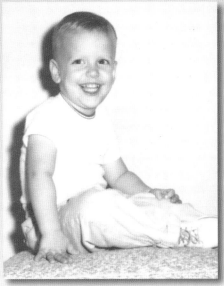

LEFT: *Silhouetted by clouds, storm chasers admire thunderheads at sunset.* PHOTO © JIM REED; WWW.JIMREEDPHOTO.COM

RIGHT: *Terry Swails perfects his television smile in 1957.*

5

REPRINTED BY PERMISSION OF JOHN L. HART FLP, AND CREATORS SYNDICATE, INC.

won the hearts of many viewers. Conrad's reports were a part of my daily routine throughout my childhood.

Another forecaster who influenced me was Tom Skilling of WGN-TV in Chicago. To this day, I'll go out of my way to catch a Tom Skilling weather cast.

My family and friends thought I would grow out of my love affair with weather. Before the field was revolutionized by the advent of computers, most children did not aspire to become meteorologists. Meteorologists were nerds, the guys with pocket protectors and thick glasses. When my friends' mothers asked me what I planned to do with my life, I could see the disappointment on their faces when I mentioned my interest in weather. I remember a conversation with a buddy regarding meteorology in which he said he hadn't known of my interest in meteors.

In high school came sports. If there was something that I enjoyed as much as the weather, it was sports, especially baseball and basketball. From an

KIT 'N' CARLYLE

REPRINTED BY PERMISSION OF LARRY WRIGHT

early age, I was successful in both. By the end of my senior year, I had enough interest from several small colleges to continue my baseball career. Still unsure about a career in weather, I decided to attend the University of Northern Iowa in hopes of making the baseball team. I figured I could get an education, play a little ball, and decipher my path in life.

It was in a Climatology 101 class at the University of Northern Iowa in the fall of 1976 that my passion for weather took flight. The professor, Dr. John Gunther, noticed I had more than a casual interest in the course. I was delighted, and relieved, to know that somebody else in the world cared as much as I did about the jet stream and its influence on weather.

One day, Dr. Gunther told me that he had just spoken with Jim Ganahl, a local TV weatherman at KWWL-TV in Waterloo, Iowa. Ganahl had told him the station was looking for someone to act as a weekend weather forecaster and substitute. Gunther suggested I call the station.

That is how I found myself in front of

Ganahl and a huge United States map on the wall at KWWL. Fronts were drawn on the map with magic markers. Magnetic temperatures and weather symbols were scattered across the map. He asked me where Minnesota was. Then St. Louis. I pointed to them. Ganahl was enthusiastic: the previous candidate had not known Iowa from Idaho and thought St. Louis was on the Ohio River.

I put on a microphone. Ganahl pointed to a red light that was on and told me to pretend to give the weather. I pointed at states, drew lines with my marker, made things up when I had to, and most importantly, I made sure to stress that St. Louis was situated squarely on the banks of the Mississippi River. I remember talking about a magnetic sun perched over Iowa that I referred to as a "fair weather friend." I still recall the bright blue cold front I meticulously drew with magic marker.

Two weeks after the audition, on a cold January day in 1977, Ganahl offered me the job. It paid $2.30 an hour, minimum wage. Three weeks later, I made my on-air debut. I was given four minutes to do the weather. I stood in front of a 10-foot-wide, hand-drawn map of the United States in the only suit I owned. I could feel sweat trickle down my arms. I was a wreck, but when the mic light clicked on, so did I. I blurted out the weather in a minute and a half. Sports had to fill the remaining time, and I recall seeing more scores, more towns, and more athletic events than I knew existed.

Years of education and thousands of shows later, I am still at it. My presentation, these days, is a technical marvel. Thanks to the arrival of computers in the early 1980s, my weather program has undergone a sea change. Magnets and magic markers have been replaced by computer-generated graphics that not only make pretty pictures, but allow me to choose from a number of graphics showing things such as temperatures, fronts, and low pressure centers. Satellites, radar, and other forms of weather information can also be displayed as single frames or animation sequences.

Terry Swails (right) and Iowa Governor Terry Branstad in the early 1980s. A long way from The Wizard of Oz!
PHOTO BY JEFF HANSEN

Sequences are created when images are stored and then assembled into a loop showing how the weather has moved over a given period of time. It was a great day for weather when the first satellite and radar images moved across the television screens. Every bit as impressive was the development of Doppler radar with its ability to perform street-level mapping and up-to-the-minute tracking.

I had come a long way from *The Wizard of Oz!*

Tornado, Hollywood-Style

Probably the most famous tornado in this country is the twister that carried Dorothy and Toto from the Gale farm in Kansas to the mystical Land of Oz in the movie, *The Wizard of Oz*. Released in 1939, the movie was ahead of its time in its use of special effects. To create the tornado that has played such a part of our collective childhood memories, the filmmaker simply rotated a 35-foot muslin windsock. While it was simple in concept, it was not cheap: this homespun twister cost $12,000 to create—a lot of money for a special effect in 1939.

7

Terry Swails catches a cat nap on the set after being named Best TV Weathercaster in 2001.
PHOTO BY BRIAN BARKLEY

One important element of my job is learning how to operate this equipment.

Stations such as KWQC, where I work, spend millions of dollars on equipment and data, which I am expected to understand. Training is an ongoing process. Once I have mastered, say, the mechanics of how to use a Doppler radar, I then have to create a forecast that is clear and concise but not complicated. If the forecast looks good but it has no substance, viewers will reach for their remote controls.

This technological revolution has made television weather much more sophisticated. Weather has become "cool," modernized, and popular, and the stigma of nerds and geeks in the profession has vanished with the wind. Today the degree of specialization is so advanced that a full-fledged meteorologist with a strong background in computers is a must for any station. The days of a kid like me walking into a job with no experience or education are long gone.

In this information age, storms can become legendary before they begin. Extreme weather is no more common now than it was a hundred years ago, it's just better documented. We now know approximately when storms are coming, how long they will last, and when they will pass. And with our higher densities of people and buildings, the storms that do strike are more likely to impact life and property.

With our newfound awareness of weather comes a fascination that I attribute to television's live coverage of storms. You can be warm and cozy in the Midwest and still watch a blizzard ravage the East Coast. You can be safely snuggled away and still feel part of the storm. This breaking weather event can be more entertaining, more dramatic than a movie.

This has created a fascination with weather like we have never seen before. Extreme weather is now sexy. It is a standard component of reality-based programming on cable outlets, such as Discovery, Fox, and the Weather Channel. Newspapers devote entire pages to the topic. A long way from *The Wizard of Oz,* movies have been devoted to weather events, including *Twister, The Perfect Storm, The Day After*

Tomorrow, and *The Weather Man,* and books such as Sebastian Junger's *The Perfect Storm* have created a hot market for weather-related disasters.

Weather is the anchor of local television stations because day in and day out, people turn to the local news to find out the weather. According to the Pew Research Center for the People and Press, weather is the number one form of news people not only look for, but demand. Talent and consulting agencies suggest 85 percent of viewers watch the local news for weather information. According to the article by John Seabrook in the April 2000 issue of *The New Yorker,* "weather events are hyped, covered, and analyzed, just like politics and sports."

It's not enough to hear about the storm, we want to experience the storm's wrath—the waves crashing, the wind whipping the trees back and forth, or the blinding snow—from the comfort of our living rooms. Doppler radar, three-dimensional graphics, and live photography make for compelling, captivating television.

It is the meteorologist who acts as the narrator in this ongoing drama, sharing insights and expertise while telling a story. In today's world, no station can control a television market without owning the weather.

Weather, a constant in our daily lives, is something we all have in common. It is always there. Always changing. It is something we want to know about, for we love its drama and its constant variety. It is also a conversation starter. After all, one of the simple pleasures of residing in the Midwest is bragging about surviving the big one.

It is the local meteorologist, like me, who tells the story. I am the link between the elements and those of viewers who battle them. Research claims that when I speak, viewers listen. I take this responsibility very seriously, because I know that people make personal and financial decisions based on my predictions. I find it rewarding that my passion for weather can improve the lives of others. I respect my viewers and value their trust.

This book is a celebration of weather. And it is a celebration of our ability to survive the worst that nature can dish out. Whether it is a tornado, flood, or raging blizzard, we all remember where we were and what we were doing at the time of the big storm. This book recounts some of the most spectacular weather events in the Quad Cities and the surrounding areas. With a combined population of 400,000, the Quad Cities is situated on the banks of the Mississippi River and is made up of Davenport and Bettendorf in Iowa, and Moline, East Moline and Rock Island in Illinois. Whether it is the Camanche Tornado of 1860 or the Duck Creek Flood of 1990 or the Armistice Day Storm of 1940, each storm is a unique story that illustrates not only the fierceness of the elements but the resilience of the people of the Midwest.

In this 1974 photo, West High's Terry Swails watches home plate intently following a lead-off single in the Iowa district baseball finals against Davenport.
COURTESY OF IOWA CITY PRESS-CITIZEN

The Weather Puzzle

Terry Swails delivers a forecast on the set of KDUB-TV in Dubuque, Iowa.

Since July 1986, I have been assembling forecasts as the chief meteorologist at KWQC-TV. One of the first television stations west of the Mississippi, the station has a long and distinguished history. It debuted in October 1949 as WOC-TV, when about 400 households in the area owned television sets. As the industry grew, so did the station. After it was sold in the late 1980s, the station's call letters were changed to KWQC. For more than twenty years the station has had more viewers than all of the other stations combined in the Quad-City market. One of its greatest achievements was in November 1991, when the media research firm Arbitron rated KWQC the number-one local news program in America.

Year in and year out, KWQC has traditionally been one of NBC's strongest local affiliates. A station does not achieve that sort of status without great people and management. Over the years, I have been fortunate to work with a group of professionals who make my job as easy and enjoyable as possible. From the boss to the anchors, producers, directors, camera operators, and everyone in between, there is an unrelenting commitment to excellence. Being a part of the KWQC family and its many accomplishments has been a dream come true. The on-air smiles between anchors go beyond the camera: we genuinely like and respect one another.

Putting together a forecast is a bit like piecing together an enormous jigsaw puzzle. I assemble the hundreds of puzzle pieces—information on temperatures, clouds, moisture, and wind—in order to put together a picture of the weather. What makes this task even more challenging is the fact that the atmosphere is three-dimensional, so instead of being flat, the puzzle looks more like a Rubik's cube.

Weather exists in horizontal and vertical layers that extend for miles above and around us. As a forecaster, my job is to take all the pieces of this three-dimensional puzzle and put them together. If I correctly interpret the interaction of winds, temperature, and moisture, my forecast is accurate.

When it comes to exceptional forecasts, I've had my share. One that comes to mind involves the Bix Road Race, an annual event in July which includes 20,000 runners and thousands of observers. For several days leading up to the race, I predicted a chance of heavy rain and thunderstorms.

The night before the race, which would commence at eight o'clock the next morning, I made a bold final call. Rain and storms would be likely during the race. The next morning, as the runners warmed up, the sky grew dark. At 8 A.M., the starter shot his pistol into the sky and at that instant, the heavens opened up. It rained cats and dogs the entire race, and I had a smug face that day!

If I misinterpret the weather puzzle, I have what is known in the business as "bust," or a forecast that is inaccurate. During my days in Waterloo, Iowa, I completed a forecast stating, emphatically, that there would not be rain for several days. Just as I walked into the weather office, the phone rang and an earnest caller told me it was pouring rain just blocks from the station. I checked the radar and saw only ground clutter, which I thought might be blocking a rain shower. I had no time to look outside before I was back on the air and, in a meltdown, I mentioned to the anchor that a heavy shower had popped up. But when I went outside, minutes later, the sky was filled with stars, and the moon was bright. I had been duped.

Forecasting the weather is a process that never really ends. Unlike a sporting event where the clock reaches zero and the game is over, one weather forecast influences another like a line of dominoes. In an atmosphere as fluid as ours, one weather event triggers another. For example, after a big storm passes, the winds switch to the northwest and the supply of moisture in the system is cut off. Skies clear. And, as air pressure rises, temperatures cool and the pattern shifts from stormy to cold and dry.

In order for me to visualize what the weather will do in the future, I need to know and understand the past. An accurate weather forecast is as much dependent on recent history and accurate data as it is on the skills of the forecaster to interpret it. My philosophy as a weathercaster is to be first, fast, and accurate. After 28 years of broadcasting the weather, I know these are the qualities that will earn me the trust and respect of my viewers.

The first thing I have to do in order to create a forecast is to consult a weather model. Weather models are essentially complex road maps of the atmosphere. To most people, a weather model looks like a sea of lines, shaded blobs, and numbers that crisscross the country—like a two-year-old's attempts at a drawing. To trained eyes, however, they are a treasure map. They depict where storms are expected to happen and how strong they will be. They are created by computers that ingest billions of bits of weather information and run them through complex mathematical algorithms every six to 12 hours. This process generates three-dimensional representations of clouds, precipitation, temperature, and wind, which can give viewers an accurate idea of what the weather will be like at any given time.

In a broad sense, weather models paint a picture using all the colors that are the weather spectrum. There are the bright, bold strokes of the jet stream; the subtle gray tones of clouds; and the indigo blues and fiery reds of the thermal pallet that represents temperatures.

As with art, however, these models are open to interpre-

Weather Folklore

Rainbow in the morning,
Sailor take warning.
Rainbow at night,
Is a sailor's delight.

Twinkle, Twinkle Little Star

What causes a star to twinkle? The answer lies in the atmosphere. When we look up at the stars in the night sky, small, shifting currents of air in the earth's atmosphere act as concave lenses or mirrors and bend the incoming starlight. When the light is bent for a short period of time, our eyes can't see the starlight. This creates the appearance of twinkling. To get past this, the world's greatest telescopes are located on mountaintops to eliminate as much of the earth's atmosphere as possible.

Terry Swails studies the Doppler radar as he prepares a forecast for KWQC television station.
PHOTO BY SEAN SMITH

Some days, a forecast comes together like a charm; some days the foundation is weak from the get-go. An example of a weak forecast would be a day when I clearly see that a front is due to arrive the next afternoon, but I can't pinpoint the exact time. If the front comes through at 3 P.M., it means less heating, less moisture and marginal instability with few, if any, thunderstorms. If it holds off until 5 P.M., the whole equation is altered by higher temperatures and moisture, which could result in violent thunderstorms with hail or even tornadoes.

Instead of presenting a clear forecast, I have to hedge and play out for the audience several possible scenarios. Thankfully, those days are few and far between, but when they happen, the forecast comes down to educated guessing. If you're not careful, a bust is waiting to happen.

There is nothing fun about blowing a forecast. People are always telling me that meteorologists have the best job in the world because they can be wrong 50 percent of the time and still get paid. People don't understand how personally I take my forecasts. I spend the day, and sometimes much of the night, trying to make a dependable forecast. In essence, I own it and am accountable for its accuracy. I know before anyone else when I've busted a forecast. Sometimes it comes down to the last hour, but usually I know of problems hours in advance. It may be as simple as the sun failing to shine or rain that wasn't supposed to fall. Other times, it is related to a change in track or a delay in speed that alters clouds, precipitation, and temperatures. The triple whammy, when all of the above happens, is the worst bust of all. It is, without a doubt, a forecaster's worst nightmare.

Any way you slice it, my forecast is my signature and something I am proud of. When I make an inaccurate assessment, I feel humiliated.

Once I know about the content of my forecast, I have to think about how I am going to present it. A meteorologist must be able to show the weather visually, as well as verbally. And I want my graphics to be attractive, informative, and

tation, and that is what makes forecasting fun for me. Each forecaster views the images differently. It is always a competition to see who "correctly" connects the dots first. For that reason, I do not look at any other forecasts until I have completed my own. I have to learn to trust my instincts, and I do not want to by influenced by my competitors.

To "dot up," as I call it, my first task in the weather office is to break down the models. I spread out the pieces of the weather puzzle and begin to construct a forecast. Every good prediction starts with a solid foundation, which for me is a thorough review of what the atmosphere has been up to. Has it been warming, getting wetter, or is there a cap preventing storms? Just like a doctor, I give the weather a check-up. From this diagnostic base, I can then build what I consider a good forecast.

Even on the dullest days, there are obvious pieces to build a base around. On those days, I know the atmosphere is stable, there is limited moisture, few clouds, and no fronts. That is a strong foundation. If it's strong enough, it will support the forecast. If not, before long it will crumble before my eyes.

Barometric Pressure Explained

Barometric pressure, simply put, is the weight of the atmosphere. By using a tool known as a barometer, we measure it in either inches or millibars. The significance of its weight is tied to the fact that cold air is more dense than warm. By measuring the weight of air, we know whether it is exerting a higher or lower force, which plays a role in the formation of clouds and precipitation. When pressures are high, air exerts a force or pressure that inhibits precipitation. When pressures are low, the air is light, buoyant, and far more favorable for rain or snow. By keeping tabs on barometric pressure, we can easily see the regions of low pressure that constitute storms on our weather charts.

easy to understand. Once again, I head to the computers. Computers with sophisticated graphics allow me the freedom and flexibility to present the weather as a story. If snow is in the forecast, I talk about where the storm currently is and a little bit about its history. Then I need to talk about when the snow will start, where the storm will go, and what will it mean for viewers—will they have to shovel or it is a glancing blow? Will it mean traffic snarls, closed roads or closed schools? A good forecast should have a beginning and end, a logical flow, and a strong story line.

Even if viewers have the sound turned off, they should be able to understand the forecast from the graphics. I determine each graphic—each snowflake, raindrop, or temperature reading. If it's going to be a scorching day, I make the sun a blazing yellow and make the temperatures a sizzling shade of red. I get to control the colors, size, and regions that best suit my audience.

While the graphics look simple to viewers, the graphics programs that create them are very complicated. The software that drives them is so powerful that an image you see on the air may consist of ten to 15 layers of weather data stacked on top of one another. The layers can be raised or lowered and even rotated 360 degrees. Each layer can be constructed to extract raw numerical data from one of the weather models and turn it into a meaningful on-air graphic. Thousands of color combinations and hundreds of symbols can be pulled out of the system. Understanding the ins and outs of the complex software and utilizing it to the fullest is a never-ending challenge.

These new computer programs have added an element of specialization that was not present in the early days of TV weather. Doing the weather is no longer a matter of writing on a chalkboard or moving around magnetic weather symbols—nobody steps in and does a weathercast without knowing the ins and outs of the computer graphics program or how to read a computer-generated weather model. Meteorologists must have a background in weather science to interpret this information for the viewers. Ten years ago, the average station might not even have a meteorologist—today an average weather staff includes two or three people and many stations are raising that number to four or more.

The sophisticated tools are also attracting people to the profession in record numbers. Doppler radar, which began appearing on television screens in the 1990s, converts radio waves to pictures to determine the location and

Long before there was a National Weather Service, people used folklore to predict the weather. Through observation, people discovered how weather affected the color of sky, the character of the wind, and the plants and animals. Most weather folklore has been around for thousands of years. Most American folk sayings about the weather originated with European settlers and Native Americans.

Weather Folklore

The moon and the weather
May change together;
But the change of the moon
Does not change the weather.
If we'd no moon at all,
And that may seem strange,
We still have weather
That's subject to change.

—ARTHUR MACHEN

intensity of precipitation. Forecasters can measure the speed and direction of the winds within these radio waves by determining the frequency change of radio waves surrounding rain or snow particles. Waves reflected by rain or snow moving away from the antenna change to a lower frequency, while those moving toward it shift to a higher frequency. A computer that is part of the Doppler uses the frequency changes to show the direction and speed of the wind.

High-resolution satellites generate pictures of storm clouds from altitudes ranging from 530 miles to 22,238 miles above ground. Taken several times an hour, these detailed images give forecasters an in-depth look at how storms are developing, moving, and intensifying. They do not show precipitation, which is why Doppler radar is

Weather Folklore

Evening red and morning gray,
Sets the traveler on his way.
Evening gray and morning red,
Brings down rain upon his head.

important, but they do show water vapor as well as cloud temperatures. Cloud temperatures are valuable because cooling cloud tops are a sign of strengthening storms.

I use the Doppler radar and high-resolution satellites for short-term forecasts, but to develop a long-range forecast takes more time. To put together a typical weathercast, I spend several hours analyzing weather models, radars, and satellites to forecast what a weather system will do. I spend another hour working on how the weathercast will appear—determining which graphics I'll use and then creating them. It is unusual for a meteorologist to create their own graphics, but I am particular about the appearance of my show. Also, I find that in the process of creating them, I can solidify my thoughts about my forecast and visualize what I will say in my upcoming broadcast.

After I've made the graphics, and the forecast is clear, I turn in a forecast to a chyron operator about a half hour before the show. The chyron operator copies the forecast into an electronic format that shows up on the screen. Then, when we are on the air, the director is responsible for making the necessary punches on the source board to bring the weather graphics up at the proper time. To aid him, I give him a written rundown of the structure of the forecast so he knows when and where things like radars, satellites, graphics, and temperatures are expected to go.

For the weather forecast to look clear, concise, and effective, the director, producer, audio technician, chyron operator, camera crew, and I must all do our jobs correctly. If any one of us fails, we all end up looking bad. In my early days at a station in Dubuque, a young director got so flustered trying to bring up a weather chart during my forecast that she punched nearly every button on the board. In a matter of 20 seconds, I was shown in front of a police station, a

A meteorologist before the days of computer-generated graphics. PHOTO COURTESY OF KWQC

black screen, a car accident, more black, a golf course, and some sports scores. I was mortified. When the weather report was over, the news anchors were laughing so hard they had tears in their eyes and wouldn't even look at me as I dragged myself back to the set from the weather wall. We couldn't get to a commercial fast enough!

Another tricky part of doing a weathercast is the chroma key. When I do a show, the weather charts appear directly behind me. In reality, there is nothing there but a green wall. As in filmmaking, the green color behind me is electronically removed and replaced with weather maps, satellites, radars or video. This process is known as chroma key. If I were to wear a lime green suit the color of our wall, my head and my hands would be the only thing visible to the viewers. Avoiding lime green suits has not been a problem.

So how do I know what I am pointing at when there is nothing behind me? That is where monitors come in handy. I have one on each side of me and directly ahead of me in the camera. I have to look into them to see where my hands are in relation to the graphics.

It takes some practice to make the whole effect look natural. Until I got the hang of it, I would find myself pointing at Chicago or some distant city instead of the one in Iowa I wanted to be pointing at. I also noticed that I was often peaking out of the corner of my eye to find a reference point. A sure way to tell a rookie is by their inability to keep eye contact. More often than not, they are looking at a monitor off to the side instead of the audience directly ahead.

In the midst of a weathercast, I am working under tight time con-

straints. News, sports, weather, and commercials all must fit into a pre-determined time (roughly 30 minutes). This time is measured out to the second, and my piece of the pie consists of three and a half minutes. The first time I tried to fill three and a half minutes it seemed to last a year, but once I overcame my nervousness, the time was up before I knew it. Now I have to work very hard to stay within my limit. If I go over my time limit, I disrupt the remainder of the show, which irritates the producers and sportscasters, and rightfully so.

On television, the edict "Thou shall not go too long" is like the eleventh commandment. To keep track of my time, I have an ear piece in which I can hear the producer telling me how much time I have left. Over the years, I've developed an internal clock that is fairly accurate so I don't pay much attention to the cues until the producer tells me I have 30 seconds left. Then I know that it's time to wrap up my weathercast.

FAMILY CIRCUS

9-2
©1998 Bil Keane, Inc.
Dist. by Cowles Synd., Inc.

"Do weathermen get paid even when they're wrong?"

COPYRIGHT BILL KEANE, INC., KING FEATURES SYNDICATE

When I'm on the air talking about the weather, I ad-lib. Anchors, and to some degree sportscasters, have a script that they read on a prompter. I deliver my weathercast on the spot. There are no mulligans or do-overs. Early on, I tried to memorize most of my weathercast. That worked for about 40 seconds—and then I would forget my script! Finally, I wised up and decided to just follow my instincts and tell a story based on the graphics. It's really not that hard when you understand what you are looking at. For example, on the weather map you see a cold front, some rain symbols, sun, and warm air

15

over the Midwest while cold air is situated over Rockies. The next step is to fill in the blanks. On the air it would sound like this: "A strong cold front will be pushing out of the Rockies spreading rain into the Midwest. That means we will lose our sunshine and warm temperatures tomorrow. You'll want to keep the umbrella handy and by all means, don't wash the car." Learning to tell a story from the graphics made all the difference for me because my weathercasts then became conversational as well as informative. And I didn't have to worry about memorizing the whole thing!

This emphasis on performance surprised me when I was new to the business. I thought the emphasis was on the weather first and the performance second, but in reality it is the opposite. The ability to be entertaining is one of the most important aspects of the business. With so many sources and channels vying for the viewer, you need to have an on-air personality and you need to stand out. If you don't, viewers will take to their remote controls and you've lost them.

The other thing I had to learn to do was to be confidant in my own presentation. In the beginning, I tried to emulate people I respected, which I could do, but my performance seemed wooden. Finally, a wise news director sat me down and said, "Terry, just be Terry." It was the best advice I ever received. When I was able to relax, to convey the weather in my own terms and in my own voice, my performance became more natural, more easy-going, and effusive As my personality and my passion for weather emerged on the set, my performance improved dramatically.

Nothing, however, tops that first on-air solo. It's kind of like your first kiss. As I stood before the camera on that February night in 1977, I can remember watching the seconds tick off the clock before I had to go on the air. I was petrified. If you are going to run, I told myself, this is the time. Then the microphone clicked on, and words came out

of my mouth. I have never been so happy to get through what amounted to three minutes of my life. When the weathercast was over, I was downright giddy with relief. I felt like I shimmered. I had just been on television. My face and voice had been seen and heard throughout eastern Iowa. Television was the driving force of my generation and now I was on the other side of the screen. It was the proudest day of my life.

Along the road, I've had moments of doubt and even fear, but over time I have come to understand that I am a bridge between the present and future. I have a gift that when used properly can make a positive difference in the lives of others. You can't ask for much more than that.

Terry Swails at the weather desk at KWQC TV. PHOTO BY JULIA OSTERHAUS

From Signal Flags to Doppler Radar: Reporting the Weather

My experience consulting weather maps and Doppler radar, or going before a camera and delivering a forecast in three and a half minutes, is a far cry from the early days of forecasting the weather. In the American colonies, the first known weather observations were taken in 1644 and 1645 by Reverend John Campanius Holm, a Swedish chaplain in a what is now the city of Wilmington, Delaware.

In time, Americans such as Ben Franklin began keeping detailed weather records of their own. Famous for his kite-in-the-storm experiment in 1752, Franklin also contributed to the understanding of weather in another way. In 1743, he was the first person to track a hurricane when his plans to study a lunar eclipse were ruined by the remnants of a hurricane. During the storm, he assumed the northeast winds that battered him in Philadelphia were also causing problems in Boston. Instead, to his surprise, his friend in Boston reported clear skies for the eclipse, and the storm didn't arrive until the next day. Puzzled, he gathered observations from both Boston and Philadelphia and correctly concluded that a storm was progressive in nature and, as such, could be tracked and predicted.

Your Thermometer Explained

When we read the temperature on a thermometer, we are actually reading the measurement of the kinetic energy of the air molecules. Warm air has more kinetic energy than cold air. When air molecules collide with a thermometer, kinetic energy is transferred to the glass and then to the molecules of mercury within the thermometer. Warm air makes the mercury molecules move faster, which causes them to spread out and push the mercury up the thermometer. Cold air, which has less energy, causes the opposite to happen, and the mercury shrinks.

President Thomas Jefferson was also a devout observer of weather. It was Jefferson, in fact, who envisioned a network of weather recorders. In 1797, he devised a plan to provide weather instruments to someone in every Virginia county so that a systematic, statewide weather record could be developed. His plan was adopted—but not until 1893 when the U.S. Weather Bureau was established in the Department of Agriculture.

In the Quad Cities area, the first weather records were taken at Fort Armstrong, an army post where the responsibility of keeping weather records was delegated to the medical corps. Since the War of 1812, the medical corps had established a system of weather observing at similar posts in order to study the effects of weather and climate on the causes of disease. Post surgeons were required to take daily readings of temperature, wind, sky conditions, and precipi-tation at 7 A.M. and at 2 and 9 P.M. This continued for another 50 years.

This weather box kite was released at Drexel Aerological Station in 1917. COURTESY OF NATIONAL OCEANIC AND ATMOSPHERIC ADMINISTRATION/ HISTORIC NATIONAL WEATHER SERVICE COLLECTION

This shielded snow gauge measured snowfall in the early 1900s. COURTESY OF NOAA/HISTORIC NWS COLLECTION

Weather Folklore

If it rains before 7:00 A.M., it will quit by 11:00 A.M.

I have known this to be true 88 percent of the time for the months of May through September in eastern Iowa. In April, it is true 75 percent of the time and in October, it is true only 63 percent of the time. —Terry Swails

The first weather record from Fort Armstrong, dated December 23, 1820, indicated that 6 inches of snow had whitened the grounds of the fort. That was followed by a cold wave that produced a temperature of 20 degrees below zero on December 24, and 28 degrees below zero on December 25. If that 28 degrees below zero was accepted as a part of our modern-day climatology, it would tie for the coldest temperature ever recorded in the Quad Cities.

One of the more detailed accounts of the medical corps focuses on the winter of 1842 to 43. During "The Long Winter," the Mississippi River was frozen for four months, and the ice reached a thickness of three feet. During the winter, there were 35 days of below-zero temperatures, and the snow was 30 inches deep in the woods. People were still crossing the river by foot as late as April 10, and 3-foot snowbanks were

A rain gauge at U.S. Smelting Company, a cooperative weather station in Midvale, Utah, in the 1930s. COURTESY OF NOAA/HISTORIC NWS COLLECTION

Up, Up and Away

To take readings of temperature, pressure, and humidity at elevations where we do not have gauges, the National Weather Service uses weather balloons called radiosondes. For 60 years, radiosondes have transmitted data back to earth by means of small radio transmitters. Because radiosondes eventually drift back to earth, each one comes with a mail bag, prepaid postage, and explicit mailing instructions for its return to the NWS. The NWS then reuses the radiosonde. However, of the 75,000 launched each year, only 20 percent are ever returned.

still observed on April 16. As late as May 5, farmers were unable to plow in places where the ground was still frozen.

By the mid-1800s, there were movements in Washington, D.C., to create a national system of weather observers. In 1849, Joseph Henry of the Smithsonian Institution proposed organizing a network of voluntary observers "to solve the problems of America's storms." Instruments and a book of data forms were supplied to volunteers, and the forms were to be filled out, sent in, and filed at the end of each month. In 1850, only one observer was located in Iowa, but by 1864 the number had increased to nine.

The nation's first weather service began on February 9, 1870, when President Ulysses S. Grant signed the bill establishing the Army Signal Service's new Division of Telegrams and Reports. At 7:30 A.M. on the morning of November 1, 1870, the Signal Service took its first observations. Twenty-four sergeants from posts around the country took simultaneous reports and sent them to the central office in Washington via the military's telegraph system. These observations were then analyzed by early meteorologists and used to make the nation's first official short-term forecast. Increase A. Lapham, of Chicago, fashioned a storm warning for observers around the Great Lakes:

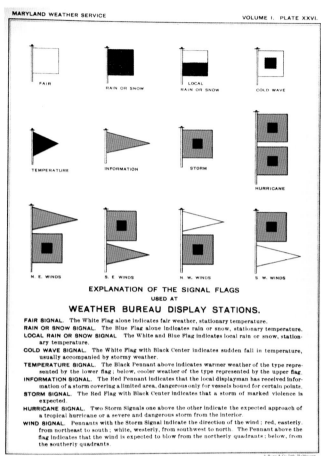

This is 1899 poster explains the various signal flags used at Weather Bureau display stations. COURTESY OF NOAA/HISTORIC NWS COLLECTION

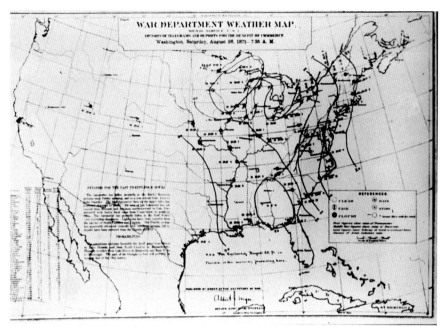

This 1871 photo features an early Signal Service weather map.
COURTESY OF NOAA/HISTORIC NWS COLLECTION

High winds all day yesterday at Cheyenne and Omaha; a very high wind this morning at Omaha; barometer falling, with high winds at Chicago, Detroit, Toledo, Cleveland, Buffalo, and Rochester; high winds probable around the Great Lakes.

In the Quad Cities, the Signal Service first started keeping precipitation records from the third floor of the First National Bank building in Davenport. They started keeping temperature records two years later, but they did not keep snowfall reports for another ten years.

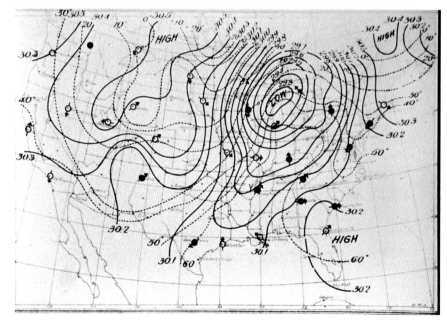

This 1889 Signal Service weather map shows a huge storm system over Lake Michigan. COURTESY OF NOAA/HISTORIC NWS COLLECTION

A shoeshine boy in 1925 uses weather as part of his pitch. COURTESY OF NOAA/HISTORIC NWS COLLECTION

By 1878, 284 posts around the nation were telegraphing weather observations three times a day to the Washingon, D.C., office for use in creating the early weather forecasts that were often called "farmers' bulletins" because they were primarily for agricultural use. In 1878, the River Stage and Flood Warning Service was created. The nation's first newspaper weather map was published in the *New York Graphic* in 1879.

The farmers' bulletins were discontinued due to an increased demand for easily accessible weather information, and in 1918 a new form of forecasting was implemented: signal flags. Instead of posting a bulletin that had to be read directly in front of the post office, a complex system of signal flags was designed to fly high above post office buildings where everyone could see them, including here in the Quad Cities.

The flags, with various shapes, colors, and sizes, indicated temperature, precipitation, and provided storm warnings. Above the other flags, the large 6-feet-by-8-feet square weather flag indicated precipitation: white meant clear or fair skies, while blue indicated rain or snow. The pennant-shaped flag was the temperature flag. If it was flown above the weather flag, it indicated warmer weather; flown below it indicated cooler temperatures. A red flag with a black center was displayed when a storm warning was in effect.

In 1891, the U.S. Weather Bureau was born when Signal Service operations were transferred from the U.S. Army to the Department of Agriculture. In 1940, the Weather Bureau's operations were transferred to the Department of Commerce. By this time, the Davenport Weather Bureau was located downtown in the Federal Building. In 1953, the Weather Bureau moved to the Moline airport, where the office would remain for 42 years.

When the National Oceanic and Atmospheric Administration was established in 1970, the Weather Bureau became the National Weather Service. Weather stations housed radar and became responsible for issuing specific storm warnings. In 1977, the Quad City forecast office was

A teletypewriter was used to transmit weather information in 1947. COURTESY OF NOAA/HISTORIC NWS COLLECTION

they still utilize volunteers to provide surface observations that are essential to improving forecasts and to providing early warnings when severe weather threatens. Every day more than 12,000 volunteers, through the Cooperative Weather Observer Program—which dates back to the days of Benjamin Franklin and Thomas Jefferson—note the temperature, precipitation, and winds in their area. These observers have also increased the base of climatological records, which play a critical role in improving long-range predictions.

The director of the National Weather Service, Jack Kelly, says these volunteers are essential. "The Weather Service couldn't get along without them," he said.

outfitted with the WSR-74C "Weather Surveillance Radar." This was a powerful, sophisticated unit run by computers. Far more detailed than previous radar, this generation of radar used color to show storm intensity while producing detailed information on storm structure.

Today, the National Weather Service's Quad Cities office is housed in one-story rectangular building on Davenport's northern edge. Staffed with 13 meteorologists, the service covers the weather for 36 counties in Iowa, Illinois, and Missouri. Along with the latest advancements in computer and satellite technology, the office is now equipped with the WSR-88D Doppler radar.

While the weather service may have the latest technology,

"Even with our sophisticated technology, satellite data, remote sensing systems, and the supercomputer, people are still the key factor in providing accurate and timely weather forecasts."

In the past two centuries, we have weathered famines, floods, and extraordinary storms. Thanks to Fort Armstrong, the Signal Service, the Weather Bureau, and countless volunteers, the rich weather history of the Quad Cities is accurately detailed and well documented.

Weather Folklore

If lightning bugs are seen flying very high during the evening, there is very little chance of rain overnight.
This is correct about 80 percent of the time.
—TERRY SWAILS

Tornado: The King of Storms

Here in the nation's breadbasket, the king of storms is the tornado. Nothing in the weather world gets treated with more respect than the dark, spinning tubes of air that come down out of the sky to rip trees from the ground, lift automobiles from the pavement, and level houses. The fickle freaks of nature are a captivating force that provoke fear and awe in even the most jaded Midwesterner.

To this day, no forecaster can say with certainty where or even why tornadoes will form. However, researchers are quick to agree that thanks to the geographical layout of the United States, this is the most tornado-prone country in the world. Cool, dry air descending from Canada gets channeled southeastward by the high terrain of the Rockies. Over the Plains and the Midwest, it encounters moist, warm air boiling up from the Gulf of Mexico. Each spring, the ensuing collision is a recipe for trouble.

It's generally estimated that less than one percent of the 100,000 thunderstorms that occur yearly in the United States breed twisters. However, when they do spin to life, they tend to form on the back edge of a rotating thunderstorm, know as a mesocyclone. Originating far above the ground, the actual tornado is born from a complex interaction of moisture, wind, and temperature.

The formation process begins when warm, humid air is drawn into the underbelly of a thunderstorm through an updraft. Once ingested, it climbs to a point where the moisture is forced to condense into raindrops. The rain-cooled air eventually sinks, producing a downdraft. Now the storm

LEFT: *A lone car shifts into reverse in the face of an oncoming tornado.*
RIGHT: *This female storm chaser photographs a dark and distant tornado.*
PHOTOS © JIM REED; WWW.JIMREEDPHOTO.COM

F-2 to F-5 Tornadoes and Tracks, 1975 to 2003

IOWA

ILLINOIS

F-3 1988
F-4 1990
F-3 1993
F-3 1992
F-3 1988
F-3 1979
F-3 1981
F-3 1998
F-3 1995
F-3 1990
F-3 1995
F-3 1978
F-3 1995
F-3 1976
F-3 1999
F-3 1975
F-4 1976
F-3 1999
F-4 1995
F-3 1975
F-3 1995

COURTESY OF NATIONAL WEATHER SERVICE, DAVENPORT, IOWA

has opposing vertical wind fields and is in a mature and strengthening stage, miles above the ground.

At this point, the final and most important part of the tornado recipe is added. That is wind shear, which means winds at the top of the storm are moving faster and in different directions than those at the bottom. This shear imparts a rolling motion that can compress the air within the thunderstorm into a spinning horizontal tube, similar to a steamroller. If wind conditions remain favorable, the tube gets tilted until one end touches the ground and a tornado whirls to life.

Tornado outbreaks have become major events for local television stations. They are hunted and tracked, with live footage that brings the drama of the chase into the homes of viewers. Standard storm coverage today includes live

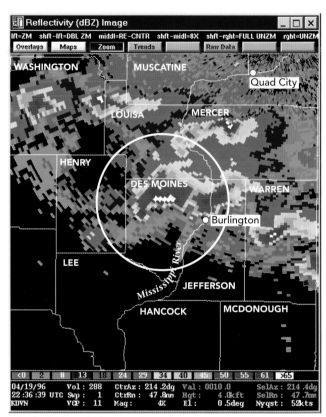

A hook echo (in circle) forms northwest of Burlington, Iowa in 1996.
COURTESY OF NATIONAL WEATHER SERVICE, DAVENPORT, IOWA

Dramatic twin funnels on April 11, 1965 in Elkhart, Indiana. COURTESY OF NOAA/ HISTORIC NWS COLLECTION

Multiple Vortex Tornadoes

"Mini-twisters" or multivortex tornadoes are composed of two or more small, intense subvortices. Subvortices are found in twisters of all sizes and can best be described as mini-tornadoes occurring within the rotation of the parent tornado. They often cause narrow, short swaths of tremendous damage that wind through the tracks of tornadoes. If a tornado doesn't contain too much dust or debris, these subvortices can be quite visible and are often misinterpreted as separate tornadoes. Most violent tornadoes are considered to be multivortex in nature.

F-2 or Greater Tornadoes in Eastern Iowa and Western Illinois, 1995-2003

Nearest Town	County(ies)	Date	Rating
Raritan, Illinois	Hancock, Henderson, Warren	May 13, 1995	F4
Washington, Iowa	Washington	May 15, 1998	F3
Hamilton, Illinois	Hancock	April 8, 1999	F3
Stockton, Illinois	Scott	May 9, 1995	F3
Mt. Carroll, Illinois	Carroll	May 9, 1995	F3
Cordova, Illinois	Rock Island, Whiteside	May 9, 1995	F3
Walnut, Illinois	Bureau	May 9, 1995	F3
Galva, Illinois	Henry	April 19, 1996	F3
Monticello, Iowa	Buchanan, Delaware, Jones	July 27, 1995	F3
Swan Creek, Illinois	McDonough, Warren	April 5, 1999	F2
Good Hope, Illinois	McDonough	April 5, 1999	F2
Covington, Iowa	Linn	July 20, 2003	F2
Cedar Rapids, Iowa	Linn	July 20, 2003	F2
Argyle, Iowa	Lee	May 10, 2003	F2
Carthage, Illinois	Hancock	May 10, 2003	F2
Seaton, Illinois	Mercer	June 18, 1998	F2
Mechanicsville, Iowa	Cedar	May 9, 1995	F2
Blue Grass, Iowa	Scott	June 14, 2001	F2
Jesup, Iowa	Buchanan	June 18, 1998	F2
Aurora, Iowa	Buchanan	September 6, 2001	F2
Atalissa, Iowa	Cedar	July 27, 1995	F2
Dixon, Iowa	Scott	July 27, 1995	F2
Grand Mound, Iowa	Clinton	July 27, 1995	F2

Courtesy of National Weather Service, Davenport, IA

25

pictures, and wind-blown reporters giving reports from the field. Doppler radars that have storm-tracking capabilities can show the audience the city streets and rivers the storm is likely to hit, while pinpointing the time of its arrival. Tornadoes are also ratings-grabbers—a station that can capture a live tornado can expect to capture the loyalty of concerned viewers.

As a meteorologist issuing the tornado warnings, I walk a fine line. On one hand, I need to get the weather information out to viewers just as fast and accurately as possible so they can protect themselves. On the other hand, I am expected to utilize my tools and resources not only to inform, but to entertain, as well.

One of the most captivating severe weather moments for

any viewer is when I rotate the Doppler sideways to show a yellow spinning tube extending from the base of a thunderstorm. The tube, known as a shear marker, indicates intense rotation within the storm that has reached, or is very close to reaching, the ground. This "low level lock" is a strong indicator that a tornado has formed or is in the process of forming. Placed over the base map of the radar, the tube offers the viewers and me a clear and dramatic three-dimensional look at exactly where the tornado is located. I can put a track on the tornado to indicate what towns are in its path and what time it will arrive, right down to the minute. I've not seen any weather coverage that is as informative and entertaining as that.

Despite being in a part of the country that is noted for twisters, most of our area has been lucky enough to avoid them the past decade. Despite the fact that tornado numbers are soaring to record levels around the country—in 2004 a record 1,717 tornadoes set down across the United States—the number of tornadoes has actually gone down around the Quad Cities.

People often tell me that they believe that the Mississippi River that flows east to west through the Quad Cities acts as some sort of tornado barrier. This is a myth. The Cordova Tornado on March 13, 1990 disproved this. This F-3 tornado became a waterspout for a time as it traveled on the Mississippi River from Le Claire to Port Byron for at least a mile. Other tornadoes have crossed rivers, producing showers of fish and frogs after sucking them out of their watery homes.

We all know that tornadoes come from thunderstorms, but the ones that cause the most death and destruction form from a special type of storm known as a mesocyclone. It's been proven that most thunderstorms self-destruct when rain-cooled air chokes off the warm updrafts that exist inside the thunderheads. However, when the right combination of wind, moisture, and heat are present, some storms (less than 10 percent) can tip slightly.

Weather Folklore

A veering wind will clear the sky. A backing wind says storms are nigh.

26

A dangerous supercell thunderstorm spins over a barren field. PHOTO BY MIKE HOLLINGSHEAD

<table>
<tr><th colspan="3">THE FUJITA CLASSIFICATION SCALE
FOR RATING TORNADO STRENGTH</th></tr>
</table>

Fujita Rating	Wind (Miles per Hour)	Estimate Typical Damage
F0	Less than 73	Light damage. Some damage to chimneys; branches broken off trees; shallow-rooted trees pushed over; sign boards damaged.
F1	73 to 112	Moderate damage. Surfaces peeled from roofs; mobile homes pushed off foundations or overturned; moving autos blown off roads.
F2	113 to 157	Considerable damage. Roofs torn from frame houses; mobile homes demolished; boxcars overturned; large trees snapped or uprooted; light objects thrown about like missiles, and cars lifted off the ground.
F3	158 to 206	Severe damage. Roofs and some walls torn from well-constructed houses; trains overturned; most trees in forests uprooted; and heavy cars lifted off of the ground and thrown.
F4	207 to 260	Devastating damage. Well-constructed houses leveled; structures with weak foundations blown some distance; cars thrown; and large missiles generated.
F5	261 to 318	Incredible damage. Strong frame houses swept off foundations and blown away; automobile-sized missiles thrown more than 109 yards; trees debarked; trains lifted from tracks; and entire towns reduced to rubble.

COURTESY OF NOAA

This tipping is a small but critical development that can keep the storm alive and well for hours. It also generates an intense rotating thunderstorm or mesocyclone. If the updraft in a mesocyclone is violent enough, the counterclockwise spin can spawn nature's deadliest storm, the tornado.

When it reaches the ground, the narrow, violently rotating column of air can spin at up to 300 miles an hour and move as fast as 70 miles per hour. Typically, the winds within a tornado are measured by the damage they leave behind. Recently however, mobile Doppler radars have been able to get close enough to tornadoes to precisely measure storm strength. In 1999, a Doppler measured wind speeds of 318 miles per hour in a twister that leveled Moore, Oklahoma.

The strongest tornado ever recorded, the Moore storm, was generated by a special class of thunderstorm known as a supercell. A supercell is a violent breed of storm that lasts for hours and exists as one giant solitary thunderstorm. They can reach heights of up to 70,000 feet and often produce a number of tornadoes over their life span. Supercells are responsible for nearly all the significant tornadoes in the United States and the majority of hail larger than golfballs.

Weather Folklore

Flowers close up before a storm.

Tornadoes come in all shapes and sizes, and strengths. Size is not necessarily a measure of a tornado's strength: a large tornado is not necessarily more powerful than one half its size. For that reason, a classification scheme was developed to rate a twister's power of destruction. Based on damage as opposed to size, the Fujita Scale ranks storms on an international scale from F-0 to F-5. The higher the ranking, the nastier the tornado.

The Fujita Scale was developed in 1971 by noted tornado expert, the late Dr. Theodore Fujita of the University of Chicago. While somewhat subjective, the scale is used by the meteorologists around the world.

Tornado Alley, which runs from Texas north through

The Tornado of May 18, 1898

The *De Witt Observer* ran this account of a family of 5 that helplessly faced the fury of the winds...

Family Tragedy: The May 18, 1898 Tornado... While the Quad Cities has avoided a direct hit by a major tornado, nearby counties have had more than their share. The worst occurred on the afternoon of May 18, 1898. A half-mile-wide tornado narrowly missed Lowden, Iowa, but killed 19 people and wreaked havoc on farmsteads in surrounding Cedar and Clinton counties.

According to an account published in *De Witt Observer*'s May 1898 edition, which was rerun in the *Clinton County Advertiser*, the Charles Flory family was killed on their farm in the hills between Goose Lake and Preston when "the cyclone apparently dropped down over the hill, and with all its power tore everything in its path. Not a vestige of anything was left

standing. Mr. Flory was out...and probably reached the barn when the cyclone swooped down on him, and in an instant he and his entire family were whirled into eternity."

When the rescuers arrived, the story said the farmer was found in a ditch just 20 yards from the house and "every vestige of clothing, with the exception of his drawers, were torn from his body, and it was fearfully bruised, nearly every bone being broken."

The article stated that the farmhouse was completed destroyed and the rest of the family was killed. Mrs. Flory was found 40 rods away, on top of a hill near the house. "Every stick of her clothing had been torn from her body," the piece went on to read. "And it had been pounded around in such an extent that the body was rolled up in a ball with scarcely a whole bone in it. Her head was down between her limbs.

The children were victims, too, of the tornado's fury. "The oldest child, Wayne, 3 years old, was found 44 rods away," the paper reported, "with a wire twisted about his neck so tight that the space covered was no larger around than a person's wrist."

This 1898 cyclone tore across western Cedar and northern Clinton counties in Iowa before crossing into Illinois. It is one of the oldest known tornado pictures. PHOTO BY G. W. TALLMAN/COURTESY OF CENTRAL COMMUNITY HISTORICAL SOCIETY, DEWITT, IOWA

A home demolished by the 1898 tornado. COURTESY OF CENTRAL COMMUNITY HISTORICAL SOCIETY, DEWITT, IOWA

The rest of the family faced a similar fate. "The twins, McKinley and Logan, 18 months, were torn piecemeal," the account read. "One was found near his mother and the other, near his father."

The family is buried together in Grand Prairie cemetery. Five small headstones and one large monument mark the resting place.

The *Tipton Conservative* carried this description of the storm on May 26, 1898:

"The cyclone of last Wednesday was a most terrifying spectacle to look upon, with its swift, revolving, black outer circle and lighter spiral center, wearing the appearance of a dense discharge of smoke from a monster locomotive. Swooping down upon the earth it would grasp buildings in its fierce embrace, and tearing it into a thousand fragments, toss them wantonly away, with a demonic roar that fairly shook the ground and filled the soul with terror. None who looked upon its awful form and irrestibly [sic] destructive force will soon forget or wish to see the like again."

Men burying livestock killed by the 1898 storm. COURTESY OF CENTRAL COMMUNITY HISTORICAL SOCIETY, DEWITT, IOWA

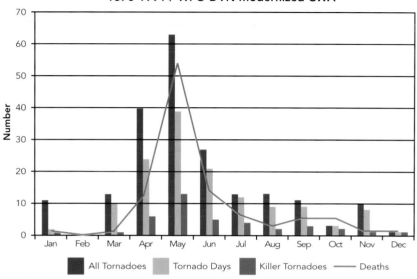

Significant Tornadoes by Month
1870-1991 / WFO DVN Modernized CWA

Legend: All Tornadoes · Tornado Days · Killer Tornadoes — Deaths

COURTESY OF NATIONAL WEATHER SERVICE, DAVENPORT, IOWA

Oklahoma and Kansas and on into western Iowa is the most famous and active breeding ground for tornadoes. Here warm, moist air from the Gulf of Mexico encounters cool, dry air from the Great Plains. This "dry line," as it is called, is a distinct moisture boundary that serves as a front. Research has proved that the dry line is potent fuel for severe weather outbreaks. The Quad Cities are not squarely in the heart of Tornado Alley, but close enough to it to be in a very vulnerable part of tornado country.

Each year approximately 1,200 tornadoes touch down in our nation, and the number of annual tornado-caused deaths is approximately 55 people. Most last from five to ten minutes, travel only a mile or two, and are considered weak. They are, on average, 75 yards wide and travel at a speed of 25 miles per hour. Violent storms (F-4 and F-5 storms) with winds up to 318 miles per hour account for just 1 percent of all tornadoes but they are responsible for 67 percent of tornado fatalities.

In Iowa, tornado season typically begins in early April and winds down in August, depending on the availability of warm, muggy air. Action can begin as early as March or as late as May. In the fall, as colder air moves into the Midwest and clashes with the last of summer's heat and humidity, tornado activity can increase again.

On average, approximately 50 tornadoes a year strike Iowa, but the number varies significantly. Even though tornadoes were especially active in 2004—with an all-time record of 120 tornadoes reported across the state—no deaths occurred. In fact, tornado fatalities in Iowa occur only about once every two years. The

Tornado Fatalities by Location

Year	Mobile	Permanent	Vehicle	Business	School/Church	Outdoors	Others	Total
2003	25	24	0	1	0	3	1	54
2002	32	15	4	0	1	3	0	55
2001	17	15	3	3	0	2	0	40
2000	28	7	4	0	0	2	0	41
1999	36	39	6	3	0	9	1	94
1998	64	46	16	1	0	3	0	130
1997	16	38	3	3	0	8	0	68
1996	16	6	3	0	0	0	1	26
1995	8	10	4	0	0	1	15	38
1994	9	11	1	0	20	2	3	46
1993	13	6	7	3	1	3	0	33
1992	20	18	0	0	0	1	0	39
1991	20	3	4	0	0	12	0	39
1990	7	11	14	15	5	1	0	53

COURTESY OF STORM PREDICTION CENTER

A fearsome mesocyclone winds up over a wheat field. PHOTO © JIM REED; WWW.JIMREEDPHOTO.COM

peak of tornado season is May and June, when more than half of Iowa's tornadoes occur, and the most likely time of the day to experience a tornado is between the hours of 5 P.M. and 9 P.M.

Thanks to great advances in research and technology, people today can monitor threatening conditions and take precautions to protect themselves if a tornado develops. Television and radio stations are required by the Federal

Communications Commission to broadcast tornado watches and warnings issued by the National Weather Service. Most communities have sirens to warn of dangerous storms. Computers, pagers, and cell phones can now receive alert signals or e-mails about tornado warnings when other media is unavailable. Weather radios can be programmed to sound an alarm when a warning is issued for a specific county. These can be lifesavers late at night, when sleep can mask the dangers of an impending storm.

Before the early tornado awareness programs of the National Weather Service, tornadoes regularly resulted in staggering death tolls. The Tri-State Tornado, for example, tore a 219-mile path of a destruction as it traveled east from Missouri through Illinois and Indiana on March 18, 1925. On the ground for three and a half hours, the F-5 storm resulted in 695 deaths, the record number of fatalities caused by a tornado. In Murphysboro, Illinois, alone, 234 people were killed, which is the highest number of tornado-related deaths sustained in a single community.

Today, meteorologists would see the conditions that produced the Tri-State storm. At the Storm Prediction Center (SPC), a branch of the National Weather Service in Norman, Oklahoma, an alert would be sent days in advance to warn the media and emergency management agencies, such as the Federal Emergency Management Agency (FEMA) and the state and local police, of a potential severe weather outbreak. The day of the tornado, prior to any severe weather, the SPC would issue a tornado watch for a specific time period. Finally, before the tornado struck, the local National Weather Service office would upgrade the watch to a warning well in advance of the actual tornado—as much as half an hour before the tornado struck. The resulting damage would still be extreme but the death toll would be hundreds less because the advance awareness and early warnings would give people extra time to find proper shelter. Since the implementation of early warning systems and Doppler radar, no single tornado has killed more than 50 people since 1971.

While detection and warning systems have improved dramatically, they only work well if people know what to do when the alarm is sounded. The supercell thunderstorm that developed about 75 miles northeast of the Quad Cities the afternoon of July 13, 2004 is a perfect example. It moved southeast, heading straight for Parsons Manufacturing, a custom metal fabrication and assembly plant, where 140 employees were hard at work.

When the plant opened in the 1970s, owner Bob Parsons had designed and implemented a severe weather

32

The Parsons Manufacturing plant after the July 13, 2004 tornado.
PHOTO COURTESY OF MATT DAYNOFF, *PEORIA JOURNAL-STAR*

plan for his employees. In the plan, a designated employee was responsible for monitoring the skies and determining when employees needed to be sent to shelters for protection. As the tornado developed northwest of the plant, the storm watcher told the employees to head for one of three reinforced shelters without windows. As employees were on their way, the National Weather Service issued its own tornado warning.

By this time, a violent F-4 tornado was bearing down directly on the plant. Just ten minutes before the tornado struck, more than 150 people had managed to get out of harm's way. The 250,000-square-foot plant was destroyed in seconds, but because employees had followed the warnings and followed the solid safety plan, not a single employee was injured.

Despite the improvements in warning technology, the

LEFT: *This 33-rpm plastic record blown into a telephone pole demonstrates the awesome power of a tornado.* COURTESY OF NOAA/HISTORIC NWS COLLECTION

BELOW: *The 2004 tornado that struck the Parsons Manufacturing plant.* PHOTO BY STEVE SMEDLEY/ BLOOMINGTON PANTAGRAPH

33

March 18, 1925 Tri-State Tornado

The Tri-State Tornado Track, courtesy "Illinois Tornadoes" by John W. Wilson and Stanley A. Changnon, Jr., Illinois State Water Survey, Urbana, IL (1971).

The tornado track of the 1925 Tri-State Tornado. COURTESY OF NOAA/ *ILLINOIS TORNADOES* BY JOHN W. WILSON AND STANLEY A. CHANGNON, JR.

Tornado Watches and Warnings

Knowing the terminology, tips and rules about what to do when a tornado strikes can save the lives of you and your family.

A tornado **watch** means that weather conditions are favorable for severe thunderstorms, which could produce tornadoes. If a watch is issued by the National Weather Service, keep an eye on the weather. If conditions worsen, be prepared to take shelter.

A tornado **warning** means that a tornado is sighted or indicated by Doppler radar. Take shelter immediately because tornadoes can form and move quickly. It is important to stay alert during severe storms.

What To Do If a Tornado Warning is Issued:

If a tornado were approaching, would you know what to do? Tornadoes are the most violent atmospheric phenomenon on the planet. Winds of 200 to 300 mph can occur with the most violent tornadoes. The following are instructions on what to do when a tornado warning has been issued for your area or whenever a tornado threatens.

- In homes or small buildings, go to the basement or to an interior room on the lowest floor, such as a closet or bathroom. Wrap yourself in overcoats or blankets to protect yourself from flying debris.
- In schools, hospitals, factories, or shopping centers, go to interior rooms or halls on the lowest floor. Stay away from glass or buildings with large roofs, such as auditoriums and warehouses. Crouch down and cover your head.
- In high-rise buildings, go to small interior rooms or halls. Stay away from exterior walls or glassy areas.
- Abandon cars or mobile homes immediately! Most deaths during tornadoes occur in cars and mobile homes. If you are in either of those locations, leave them and go to a substantial structure or designated tornado shelter.
- If no suitable structure is nearby, lie flat in the nearest ditch or depression and use your hands to cover your head.

THE WORST TORNADO OUTBREAK IN U.S. HISTORY ON APRIL 3 AND 4, 1974

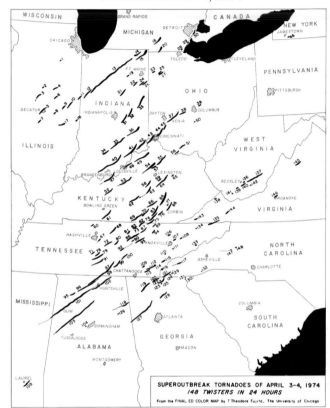

COURTESY OF NOAA

TOP 5 DEADLIEST U.S. TORNADOES

Date	Place	Deaths
March 18, 1925	Tri-State (MO, IL, IN)	689
May 6, 1840	Natchez, MS	317
May 27, 1896	St. Louis, MO	255
April 5, 1936	Tupelo, MS	216
April 6, 1936	Gainesville, GA	203
April 9, 1947	Woodward, OK	181

COURTESY OF NOAA

The April 20, 2004 tornado that killed 8 people in Utica, Illinois. COURTESY OF JAMES KRANCIC

phenomena known as Super Outbreak on April 3 to 4, 1974 resulted in a number of deaths and a great deal of destruction. In the afternoon and evening of those early spring days, 148 tornadoes battered the central United States and a small part of southern Canada in 24 hours. Three major squall lines developed that killed 315 people, destroyed more than 600 square miles of anything in its path, and resulted in more than $500 million in damage. Eleven different states reported fatalities, with the most occurring in the town of Xenia, Ohio, where 34 deaths were attributed to one extremely violent F-5 tornado. If the outbreak had occurred 25 years earlier, some experts felt the death toll from the Super Outbreak may have climbed above 1,000.

Imagine what it must have been like back on May 9,

1918 when one of the most powerful twisters in Quad City history roared into Eldridge, Iowa, at 6:35 P.M. The only warning that the residents had was word of mouth. Here is a recounting of the tornado that appeared in the May 10, 1957 edition of the *Democrat Times.*

Before the storm, all was peaceful and serene in Eldridge. The sun shone after a brief rainstorm. A rainbow spread itself over the heavens.

A vast gathering of clouds was then seen on the horizon. They assumed an alarming stage of formation.

Veteran weather forecasters at once observed them as a cyclone in the first stages.

"A cyclone, a cyclone!" shouted residents of the village. The warning was quickly carried from house to house until all had been warned.

Mrs. Ben Litscher, then 16-year-old Emma Damann, remembered the storm vividly.

We knew a storm was coming but we couldn't see the sky to the west too well because of the trees in a schoolyard nearby. We didn't pay any attention until we heard a loud roaring noise.

Even then we thought it was just the trains switching on the track east of the house. But when we looked out the windows and didn't see any trains we knew something was wrong.

The first thing we saw were a lot of things flying through the air, pieces of trees and houses. Then my mother and I tried to get into the basement but the wind pressure was so strong we couldn't get the door open.

When we tried to open the outside door, the wind held it so tightly we couldn't move it. Glass in the windows started to break from the pressure and drove us back. Finally my mother got outside but I didn't make it.

I remember trying so hard to find something in the house to hold on to. Everything was moving and then suddenly I didn't remember anything. When I came to I was in a neighbor's house and they were calling the doctor.

The May 9, 1918 twister that struck Eldridge, Iowa, injured 18 people and levelled 8 homes. COURTESY OF *QUAD-CITY TIMES*

Mrs. Litscher had been found in a neighbor's yard a block away. Her collarbone was broken in several places, a number of teeth were knocked out, her knee was badly wrenched, and she had a deep cut on the shoulder where she had been struck by some flying object. Her home was blown away.

Another mesocyclone hovers over the Midwest. PHOTO BY MIKE HOLLINGSHEAD

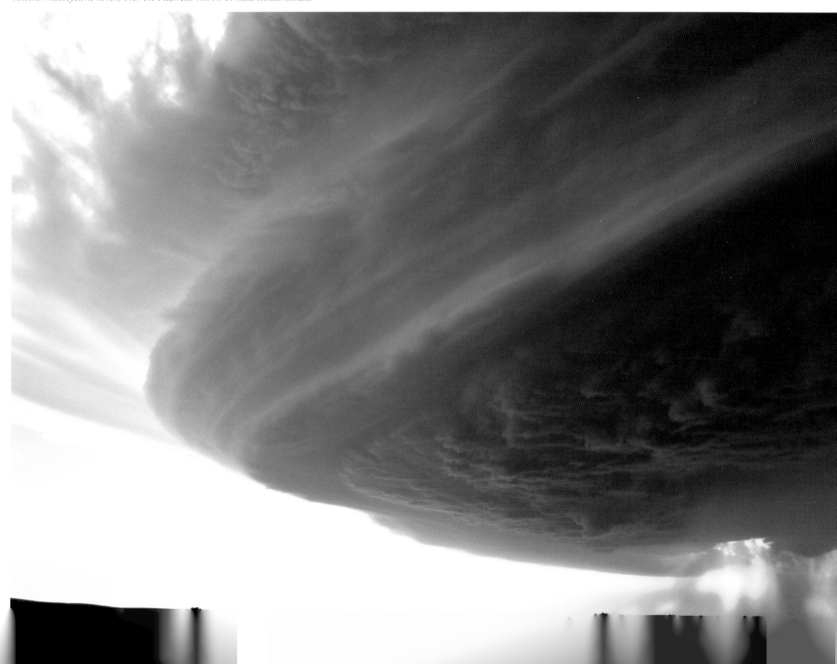

Camanche Tornado of 1860

The early settlers in this area knew little about tornadoes except that they were dangerous and they came without warning. Before the days of Doppler radar, tornado sirens, and live television coverage, the intensity of a storm was rarely known until it was on top of those in its path. To survive them, people needed common sense and good luck. Sometimes even those weren't enough.

The Camanche Tornado of 1860 was dangerous, unannounced, and deadly. In four hours it traveled east from Lowden to Camanche, Iowa—just 25 miles north of the Quad Cities—and left a trail of death and destruction. The violent twister crossed the Mississippi River, leveled Albany, Illinois, and headed to Lake Michigan, just north of Chicago.

The Camanche Tornado is the most severe storm that has ever struck this area. Traveling at speeds of up to 66 miles per hour with winds of up to 300 miles per hour, the tornado left 141 people dead and 329 injured and scorched a trail of complete destruction about half a mile wide. Many of the injured died later, bringing the death toll to a staggering 200 people. It was front-page news throughout the world.

John L. Stanford, professor of physics at Iowa State University, compiled a graphic and comprehensive account of the Camanche Tornado in his 1977 book, *Tornado.*

The year before the Civil War, settlers in eastern Iowa were enjoying a leisurely Sunday afternoon in early summer. The usual weekday bustle and activity was absent from the streets of the little town of Camanche, founded two years earlier on the west bank of the majestic Mississippi River. The market and the brick schoolhouse were deserted. The town of 1,200 persons relaxed on this day of rest, June 3, 1860. At midafternoon, the darkening western sky suggested that rain might cool off the oppressive heat of the day....

Stanford said the storm formed in western Iowa and intensified as it moved into central Iowa, "dragging a trunk-like tornado on the ground." Then it began to hit the small communities in its path.

In Hardin County, it destroyed the small town of New Providence. Fortunately, most of the townspeople were some distance away attending an afternoon Quaker meeting. The tornado struck the village from the northwest, wrecking 11 houses and leaving several persons injured. It next hit the tiny community of Pritchard's Grove, 6 miles to the east. Here the townspeople were not as fortunate: seven persons died as the tornado became a killer storm. One witness related that "timber and every movable thing were swept away like dust before a broom."

As the storm moved east, Stanford said, its "violent cyclonic winds" flattened everything in its path.

Nothing was left above the ground. Only cellar holes remained to show where houses had formerly stood. Many

A Harper's Weekly *artist drew this rendering of the coffins of Camanche residents killed by the 1860 tornado.* COURTESY OF CAMANCHE HISTORICAL SOCIETY

serious injuries occurred here, but no fatalities....

By the time the storm system reached the Cedar Rapids area, two distinct tornadoes could be seen simultaneously, one about 12 miles north of the other. One passed to the north of Cedar Rapids while the other passed to the south.

Historian Benjamin F. Gue, the lieutenant governor of Iowa from 1866 to 1868, was 31 years old in 1860 and lived near Cedar Rapids, close to the town of Mount Vernon, when he observed this tremendous tornado at close range as it passed by only half a mile from his farm home. In his four-volume *History of Iowa*, published in 1903, he recalled how he and his family watched the storm clouds rushing toward them:

We could now faintly hear long continued rumbles of thunder and for some time sharp tongues of lightning had been visible. The atmosphere, the haze and the rising bank of clouds had a weird unnatural appearance and the oppressiveness of the lifeless heat became almost unbearable.

It was now noticed for the first time that the light-colored upper clouds, which resembled the dense smoke of a great prairie fire, were rapidly moving from the north and south toward the center of the storm cloud, and as they met, were violently agitated like boiling water descending in a rapid movement to the black cloud below. We were all now intently watching this strange movement, something we had never before seen, when the thought flashed across my mind—this is a tornado!

Gue described the formation of the tornado, the clouds rising rapidly to form an immense funnel "the lower end of which seemed to be dragging on the ground." The sound, Gue said, "was a steady roar, very heavy but not loud, like an immense freight train crossing a bridge." Then, as the tornado began moving, it began to tear its way across the landscape.

We saw high up in the air great trees, torn and shattered, thrown by the force of the whirlwind outside of its vortex and

An artist's rendering of the 1860 tornado that killed 29 Camanche residents.

falling towards the earth. My family had gone into the cellar, which was of large rocks, upon which rested the balloon frame house. I stood close by the outside doorway ready to spring in if the fearful black swaying trail should come toward the house. It appeared to be passing about half a mile north of us. The sight, while grand and fearful, was too fascinating to be lost unless the danger became imminent.

The roar was now awful, and a terrific wind was blowing directly toward the swaying, twisting black trail, which seemed to be sweeping down on the ground. It was coming directly toward the log house of my nearest neighbor on the north, and I saw the family run out and down a steep bluff of Rock Creek and cling to the willows. Suddenly the funnel rose into the air and I could see falling to the earth, tree tops, rails, boards, posts and every conceivable broken fragment of wrecked buildings...No pen or tongue can convey to the mind a true picture of the frightful sights and sounds

Weather Folklore

Expect rain and severe weather when dogs eat grass.

HARPER'S WEEKLY. [JUNE 23, 1860.

This Harper's Weekly rendering illustrates the Camanche Tornado's devastation: only 20 buildings were still on their foundations.
COURTESY OF CAMANCHE HISTORICAL SOCIETY

THE GREAT TORNADO.—RUINS OF ALBANY CITY, ILLINOIS.

that lurked in the rear of that irresistible tornado as it was then gathering greater power of destruction to overwhelm and crush the town of Camanche.

The editor of the *Mt. Vernon News* also witnessed the event, which he described on June 4, 1860:

When first seen, probably six or seven miles away, it had the appearance of a large black shaft or column…Hundreds watched as it swept on its course, seemingly bearing directly on Mt. Vernon.

When within two miles of us, while people were seeking safety in cellars, or, in some cases, running wildly about the streets, it veered on its course, and swept by in full sight—sublime, but fearful. Hardly had it passed ere a half-dressed man bleeding from wounds on his head, and reeling upon his horse, rode furiously into town, calling for help. Talking incoherently, he reported persons killed and others injured at a little village or hamlet one and a half miles west, known as St. Mary's.

The storm continued its way east through the many small towns strung along the railroad line that is now U.S.

Route 30. It smashed a train depot, warehouse, and ten loaded freight cars at Lisbon. The east-southeast path of the storm, roughly paralleling the railroad, added to the destruction. By the time the storm had reached Clinton County, it had already claimed at least 42 lives in the sparsely settled interior of Iowa.

The two tornadoes were 7 miles apart as they crossed into Clinton County. The northernmost tornado struck Lisbon, then joined its southern twin at Wheatland and DeWitt. The southern tornado was first observed 7 miles southwest of Cedar Rapids, and it was the tornado Gue so graphically described.

The two tornadoes had coalesced into a single, twisting giant; then the whole storm rushed east, leaving a strip like a desert from a quarter-mile to a half-mile wide. As the tornado passed between DeWitt and Camanche, an additional 28 persons died.

From their farm 3 miles west of Camanche, the Ralston family saw a huge, black, funnel-shaped cloud extending down from the angry clouds overhead. As the whirling tornado approached their house, they ran in desperation to seek shelter in a grove of locust trees. The family fell to the ground, literally hanging on for dear life to the tree trunks. The tornado swept across their farm, lifted their house, and carried it about 100 yards to the west. Then it whirled the house back almost over its original location and shattered it to pieces. As Ralston said, the house was "rubbed out as you would rub a snowball between your hands." The cloud of murky blackness then swept on toward Camanche, obliterating everything in its path.

A correspondent of the *New York Herald* happened to be traveling toward Camanche as the tornado struck. He wrote these lines from Clinton on June 5, 1860, two days later:

I was but a short distance from the village at this time, and was an unwilling witness of the heartrending scene. No

40

pen can adequately describe the scene of terror, agony, and peril that ensued. The air, darkened by the immense moving cloud, charged with death, the rain, which was now falling in torrents, the fragments of crushed and scattered buildings, which were flying in all directions, and the shrieks and groans and prayers for help that were heard, even above the din and roar of the tempest, all combined, rendered the scene one of the most solemn and extraordinary I ever witnessed…The angel of destruction has passed over it, and with his wings had brushed it from the bosom of the plain.

In the *Lyons City Advocate* of June 4, 1860, a reporter described Camanche immediately after the tornado:

We found the town, as the messenger has reported, literally blown to pieces, and destruction and death scattered everywhere…The first pile that met our eye was the ruins of the Millard House…a three-story brick hotel [that] could not have been more effectively destroyed had a barrel of gunpowder been exploded within its walls.…

Hardly a house was left uninjured, and that many of them were swept entirely away. Every business-building in the place is destroyed, including a large brick block recently erected. About sixty feet of this, including the cupola, is demolished, and the remainder is nearly unroofed…The dwelling and store of Mr. Waldorf, a three story brick, is entirely demolished, and the family buried in the ruins. Mrs. Waldorf and one child were taken out dead, and two children rescued alive—and strange to say—unhurt. Mr. Waldorf had not been found when we left, at 2 o'clock A.M.

Another reporter, published in the *Vincent's Register*, portrayed the human side of the disaster:

Dr. Howell of Fulton informed us that from Sunday evening to Monday noon he had visited ninety-one wounded, and set twenty-three broken limbs.

We saw twenty-eight dead bodies, and there were eighty-two that required strict medical attendance, and as many more that are more or less hurt, yet are able to be around. In addition to this number, twenty-eight were swept from a raft that was passing at the time, and ten are yet missing from the town, that are supposed to be buried in the ruins or have been blown into the river and drowned.

In less than three minutes, the monster funnel had

The Legend of Low Moor

COURTESY OF CLINTON HERALD

There's gold in Low Moor

According to legend, a pioneer family was headed to the new frontier from Illinois when they encountered the Camanche Tornado 6 miles west of Low Moor, located just north of the Quad Cities in Iowa. Their belongings in a covered wagon included a buckskin sack full of $7,000 in gold coins, which they intended to use on a land purchase.

When the tornado hit, every member of the family was killed except for a little girl. Reportedly, the wagon and the sack of gold were lifted hundreds of feet into the air and then smashed against a rail fence, scattering the gold coins in all directions.

In the wake of the tornado, survivors collected $3,000 worth of the coins. The other $4,000 were nowhere to be found. So, the legend states, somewhere in the fields around Low Moor, $4,000 in gold coins await a finder.

Many have shrugged off the story as nothing more than a tall tale but on several occasions gold coins have been found. Most have surfaced on the Truelsem farm, adjacent to where the pioneer family had stopped. In the past 50 years, at least four gold pieces have been found, all in good condition with the minting date 1858 clearly visible.

Derecho

Derecho is a Spanish word that means "straight ahead." It is also a term that meteorologists use to describe windstorms that accompany large thunderstorms. These winds can be long-lived and very destructive, reaching speeds of more than 100 miles per hour as they move through squall lines. The damage produced by a derecho is different than one of a typical tornado because it scatters debris in a specific pattern, rather than in all directions.

passed through Camanche and crossed the Mississippi River into Illinois. The town of Albany, Illinois, was demolished, and five persons were killed. At about 9 P.M. the tornado passed south of Dixon, Illinois. Six hours after it had first been sighted, the tornado was as far east as Amboy, Illinois—about 48 miles from Camanche.

As soon as the tornado left Camanche, darkness fell. A messenger was dispatched to nearby Clinton, 6 miles to the northeast along the Mississippi. At Clinton, the rain was over. The air was balmy, and a few stars peeked through holes in the night clouds. The rider from Camanche dashed through the streets calling out, "Camanche is destroyed by a tornado, and half the inhabitants are buried in the ruins! Send down all your doctors and materials to dress the wounded!"

In the *Palimpest* in April 1933, an unidentified writer described the scene he had witnessed nearly 73 years before:

Volunteer workers picked their way over fragments of buildings, fences and loose materials of all kinds to the few shattered fragments of houses that still remained upon First Street…Parents were weeping for their children and children for their parents. Here a husband bent sobbing over his dying wife, and here a mother, with frantic joy, pressed to her bosom the child she thought was lost and found to be alive. Many seemed blessed with a calmness from on high; many were beside themselves and many were bewildered and overcome with stupor.

Hour after hour they worked frantically. The ruins strew around, the hideous distortions of the dead, the mangled bodies of the living, the multitudes of eager, grimy workmen, the peaceful summer night and the clear moonlight overhead, formed a scene never to be erased from the minds of any who were present.

The power of the tornado was incredible. Although it lasted only two or three minutes at Camanche, that village of 1,200 persons was almost totally destroyed. According to Gue, "It is estimated by several meteorologists who made careful investigation to ascertain the velocity of the circular motion of the winds which wrought such fearful destruction, that it was at the rate of about 300 mph."

Amidst their shattered surroundings, survivors in Camanche were forced to plan funerals for loved ones lost to the storm. By the following Tuesday, 25 coffins were neatly arranged in front of Dunning's Bank for a mass funeral. *The History of Clinton County* contains this account: "Over two thousand sympathizing friends and neighbors were present and frequent outbursts of grief amid the deep hush that pervaded the assemblage attested to the profound grief of the stalwart men as well as the tender-hearted women."

For years afterward, old-timers would tell about this storm. Many of these stories were no doubt exaggerated, but the Camanche Tornado, as it became known, was one of the most severe tornadoes experienced by settlers of the region known as the Old Northwest.

Weather Folklore

Horses run fast before a violent storm or before windy conditions.

From Twister to Blizzard in 48 Hours

On January 24, 1967, American troops were entering the Mekong Delta in Vietnam. President Johnson warned of an impending recession. The mood of Quad Citians was about as gray as the January sky, and the forecast wasn't helping. An arctic front was winging its way through the heartland, and the local weather wizards warned that winter was about to tighten its grip.

Cold air descended on the morning of January 18. The temperature in the Quad Cities hovered at 11 degrees below zero and the region shivered in the midst of a deep freeze. A few days later, as luck would have it, a south wind sent temperatures soaring into the 50s. Winter skipped town, coats disappeared, and smiles were plentiful. For four days, the spring-like weather remained.

Then, on January 24, 1967, the weather charts started trickling in, making little more than a hum as they rolled off the facsimile machine. Silent as they were, they portrayed a merger of remarkable wind and moisture that would impress meteorologists for years to come.

A low pressure system was swirling over the Kansas

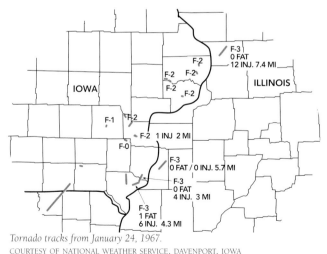

Tornado tracks from January 24, 1967.
COURTESY OF NATIONAL WEATHER SERVICE, DAVENPORT, IOWA

A tornado on January 24, 1967. PHOTO BY FLOYD REIF,
COURTESY OF JOHN STANFORD, *IOWA TORNADOES*

43

plains. The system grew stronger as it pulled into its core the spring-like warmth and moisture that flowed northward out of the Gulf of Mexico. To the northwest, the back side of this same system gobbled up the frigid polar air that blew in from Canada. When the two air masses collided, the storm literally exploded. As it spun toward Iowa, air pressure tumbled.

The Quad Cities basked in the warmth of a day more typical of April—temperatures climbed into the 60s, unprecedented for January in Iowa. In northwest Iowa, however, temperatures slipped into the teens as the cold Canadian air barreled into the state. Where the two air masses converged, massive thunderstorms sprang to life.

Just as unusual as the January thunderstorms was the position of the jet stream. Far above, the jet stream's powerful winds flowed over the surface winds in a way that caused strong directional change. In other words, winds at the ground level blew from the south, while winds aloft were from the west, imparting the spin that is ideal for creating rotating thunderstorms. All the ingredients for a tornado

The January 24, 1967 morning pressure plot showed a deepening storm over Kansas. COURTESY OF NATIONAL WEATHER SERVICE, DAVENPORT, IOWA

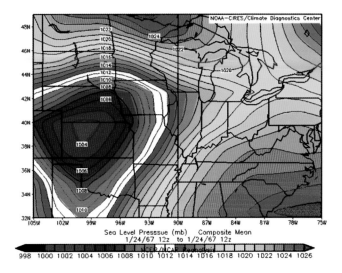

were present, and supercells began to form. In just hours, the first and only twisters in January history would churn through the barren Iowa fields.

In weather stations southwest of the Quad Cities, radarscopes began to pulse. Throughout the area, teletypes were furiously spitting out the message that dangerous storms were on the rampage. By the time a violent F-4 tornado sprang out of Missouri into southeast Iowa, at least seven other twisters had touched ground in Missouri. As the afternoon progressed, they headed across southeast Iowa directly toward the Quad Cities.

Finally, at 4:25 P.M., Helen McKinney, a storm spotter from Wapello, Iowa, dialed the number of the Quad-City weather bureau. Her words to the office were simple: a funnel was on the ground and it was headed north. By 4:30 P.M., as lightning flickered and the winter skies turned an ominous shade of green, the weather service had issued a tornado warning for the Quad Cities.

Weather Folklore

Ants are busy, gnats bite, crickets sing louder than usual, spiders come down from their webs, and flies gather in houses just before rain or severe storms.

To the south, the skies erupted. At 4:30 P.M., west of Wapello, Mrs. Dean Walker and a friend huddled in her basement as a tornado destroyed her house. West of Muscatine, Mrs. Robert Fridley watched as the tornado peeled the roof off her brick farmhouse like a potato peeler might peel a potato. Shortly after 5 P.M., Mrs. Wesley Rock of Eldridge, Iowa, looked out her window to see a tornado chewing up an open field. In seconds, it demolished a neighbor's barn and garage but spared the home.

On a relentless track northeast, the storms zigzagged through Wheatland and Clinton, Iowa, before producing an F-3 tornado that injured 12 in Mount Carroll, Illinois. Then—as fast as they came—the tornadoes were gone.

By the time the storm was over, a swarm of 32 different tornadoes had touched down in the Midwest. The first occurred in Missouri at 11:50 A.M. and, nearly nine hours later, the last one was in Wisconsin at 8:40 P.M. Twelve were classified as strong to violent (F-2 to F-4). And this was the first time in recorded history that tornadoes had occured in Iowa during the month of January.

Damage was widespread. Throughout the Midwest, 2,000 families suffered losses estimated at $15 million, 168 homes were destroyed, and another 258 homes were heavily damaged. The closest a tornado came to the Quad Cities was Eldridge, 5 miles to the northwest, but the area did experience damaging winds of up to 60 miles per hour.

Seven persons were killed in the outbreak: five in Missouri, one in Iowa, and one in Illinois. The worst news came from rural Fort Madison, in eastern Iowa, where a three-year-old boy, Byron Swyler, perished in a demolished schoolhouse.

Before residents began to pick up the pieces on that January day, winter returned. In a few hours, temperature plummeted 30 degrees, and a day that had started in the 60s—with the promise of spring—shivered to a violent and icy close.

As emergency crews worked to restore power that evening to more than a 1,000 homes around the Quad Cities, weather charts were quickly revealing a new story. The front, which had produced tornadoes and dropped temperatures, was coming to a screeching halt. A secondary low pressure system was developing over the plains of Oklahoma however, and promised to be every bit as strong as the first.

This time the Quad Cities would be on the cold side of the storm and the precipitation that was sure to fall would come in the form of snow. With near-record warmth and ample moisture lurking just to the south, the stage was set for the rapid intensification of the new cyclone. If the weather charts were right, a significant snowstorm headed toward the Quad Cities.

The morning of January 26 dawned cold and gray in the Quad Cities. Temperatures fell into the 20s. The barometric air pressure fell throughout the morning, and winds veered to the northeast. The old-timers say that any east wind is an ill wind, and in this case, they were right. As the storm deepened over Arkansas, the clouds drew closer to the ground and a few spits of snow drifted from the sky.

By evening, less than 24 hours after the tornadoes, a snowstorm was in full gear. People throughout southeast Iowa stood outside, staring in disbelief, as snowflakes swirled frantically to the ground. Snow fell steadily throughout the night, and by noon the next day, was 8 inches deep. As the center of the storm rolled across Missouri and Illinois, it took aim on Indiana with 30-mile-per-hour winds and drifting snow. High winds and enormous snowfall produced widespread blizzard conditions. Nine hours later, the last flake hit the ground.

The storm is still considered one of the worst of the century for much of the Midwest. It snowed continuously for 26 hours, and a record 13 inches of snow fell in the Quad Cities in that amount of time. Other record accumulations of snow included the 13.5 inches that fell in 24 hours on Burlington, Iowa. Farther east, the storms buried Chicago with an all-time record of 23 inches of snow and produced 60 fatalities.

The January tornado outbreak and ensuing blizzard were incredible storms in and of themselves. But the fact that

Chicago's Calumet Expressway was paralyzed by the January 26 blizzard. COURTESY OF NOAA/HISTORIC NWS COLLECTION

these storms came back-to-back in just 48 hours makes this some of the most remarkable weather in Quad-City history. This extraordinary occurrence—with the extreme variations in weather conditions—was one of the most remarkable 48 hours of weather in Quad City history.

The 1990 Tornado Outbreak: Live!

The afternoon of March 13, 1990, I stood face to face with a real-life twister as I broadcast live from the studio of KWQC-TV. It was an experience made possible by a camera mounted high above our television station. As the twister spun to life, I was able to stand next to an image of the tornado and warn people that, indeed, it was real and potentially deadly. In the most surreal moment of my life, I stood, in the calm of the studio, listening to the beating of my heart, as just miles away all hell was breaking loose.

On one hand, it was a dream come true. Only a handful of broadcasters (including Dan Rather) have ever been able to astonish their audience with a real-life tornado. On the other hand, it was my worst nightmare. Here was a potential disaster—a tornado in the midst of a major metropolitan area. The stakes had never been higher.

It is one thing to talk about a tornado; it is another to talk about it while it is spinning over your city. Nothing in my 28 years of weather broadcasting has come close to top-

A picturesque tornado descends on a field of corn.
PHOTO © JIM REED;
WWW.JIMREEDPHOTO.COM

ping the experience of tracking this harrowing storm. More than a decade later, people still remember the fear and urgency conveyed by those haunting images of the massive black clouds.

I distinctly remember looking at the weather charts on the afternoon of March 7, 1990, with a degree of doubt. Spring seemed a long way away. The temperature outside measured 36 degrees and a brisk east wind swept the cloud-stained sky. The weather charts sprawled on my desk, however, told me that in a matter of days, the winds of the jet stream would change and blow the chill from our area. If all went well, our temperatures would climb from the 30s into the 70s and, possibly, we would see some record highs—a possibility that cheers even the most jaded forecaster.

Then I noticed the moisture the warmth would bring. As temperatures and humidity increased, thunderstorms became likely. Even though, normally, the severe weather season was weeks away, what I saw spelled out before me on the weather charts was the potential for severe thunderstorms.

Before I left work that day, I made a note to review the severe weather procedures and to get my staff and my equipment into severe weather mode. This was an annual ritual, so it had to be done sooner or later, but this was a little sooner than I preferred.

I also arranged a meeting with Civil Defense Director Bud Whitfield. The Civil Defense Agency had contacts with spotters and law enforcement officials and was a vital link in providing adequate warnings in times of severe weather. I wanted to make sure the lines of communication were open between us. We arranged a meeting for the afternoon of March 13—five days away.

The next day, I reviewed the station's severe weather policy, which I had helped design. In order to issue a warning, I had to determine the urgency assigned to specific weather conditions. Tornadoes were always at the top of my list. When tornado warnings are issued by the weather service, they get full and instant attention. The meteorologist goes

The March 13, 1990 tornado on the edge of Port Byron, Illinois, became a waterspout when it touched down on the Mississippi River. PHOTO BY HANS THORNBLOOM/COURTESY OF CHUCK YOUNG

47

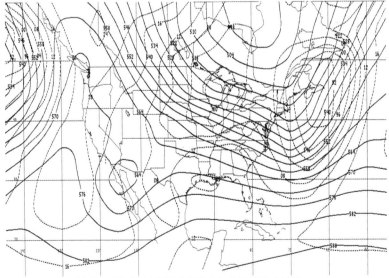

WSI SUPERFAX(TM) DATA FROM THU 3 MAR 05: 00Z MRF MODEL
144 HR FCST 500 MB HEIGHTS (M) VALID 00Z WED 9 MAR 05 ———
144 HR FCST SURFACE PRESSURE(MB) VALID 00Z WED 9 MAR 05 ------

A 500 millibar chart that provides necessary data for predicting severe weather outbreaks. COURTESY OF WSI CORPORATION

48

This incredible sequence shows the March 13, 1990 tornado that formed between Le Claire, Iowa, and Port Byron, Illinois. Note the water being sucked off the top of the Mississippi. PHOTOS BY CHUCK YOUNG

on the air and stays on until the situation is under control. Speed and accurate information are critical. Everyone involved in the process is expected to perform at his or her best. Failure to do so could mean the loss of a job. Even worse, it could mean the loss of a life.

When I dug into the latest surface plots, I noticed 80-degree warmth was surging out of Texas on the jet stream. Thunderstorms were already popping in the Lone Star State, and I was confidant they would find their way northward. In the meantime, the temperature in the Quad Cities had already increased 19 degrees over the previous day's high. It was 55 degrees and, as the saying goes, the heat was on!

I spent Friday, March 9, observing and analyzing maps, computer charts, and models that indicated a potent swirl of energy entering the western United States. This swirling, known as vorticity, was dropping into a pocket of cool air over the Rockies called a trough. From there, it would set two things into motion. First, a deep low pressure center would form over Colorado, strengthening the jet stream. This powerful ribbon of wind, which extends around the globe, would develop a branch that would flow from Texas to Wisconsin, transporting unseasonably warm, moist air into the middle of the country.

The second development would be a cold front that would trail from the low pressure center in Colorado. As the cold front arced into the warmth and moisture over the Midwest, it would ignite powerful thunderstorms. With plenty of wind shear, I could see mounting evidence for rotating storms—the kind that produce tornadoes. At this point, I realized that a major tornado outbreak early in the season was likely for the Midwest.

As I left the station that night, temperatures were still hugging 50, and my winter coat was on my arm instead of my back. While I was looking forward to the warmth that weekend, the gathering storm would never be far from my mind.

When I pulled into the station that Monday afternoon, March 12, my face was sunburned. Sunny skies and the 74

degrees had brought me and everyone else out Sunday for a dose of warm air. Throughout the Quad Cities, the smell of charcoal and hamburgers drifted on the breeze. Kids pedaled their bikes on the sidewalks with abandon, enjoying the unseasonable warmth.

Trudging up the parking lot steps, I noticed that in spite of the blazing sun and warm temperatures, a south wind was blowing. Gusting at clips of up to 50 miles per hour, it was a sure sign that the atmosphere was in an aggressive and highly unbalanced state. The wind wrapped my tie over my shoulder and made my hair stand on end in the few minutes it took me to reach the front door.

As the mercury soared, I holed up in the weather center to sort out the pieces of the day's weather charts. By late afternoon, the thermometer climbed to 80 degrees. The National Weather Service issued a special weather statement stating that the 80-degree temperature was not only a record for March 12, it was the warmest temperature ever for that early in the season. As I entered the record in my daily log, I noticed a tornado watch was issued for our neighbors to the west. The storm was on the move out west and their troubles today would be ours tomorrow.

The news on the weather charts for the next day was unmistakable: the conditions that produce violent thunder-

This dramatic view of the March 13, 1990 tornado was taken near Le Claire, Iowa.
PHOTOS BY DAVID VOSS

storms would be coming together over much of the Midwest. The risk of hail, high wind, and tornadoes was significant. The Storm Prediction Center was confident enough to issue a rare nationwide statement that read: "This is a particularly dangerous situation with the threat of large and damaging tornadoes."

Adrenaline shot through me. I made sure to inform the team I would rely on—the directors, photographers, and newsroom staff—about the possibility of a severe storm. If things got nasty and warnings were issued, I wanted their reports to be timely and accurate. When lives are at stake, no detail is too small to ignore, right down to making sure there is paper in the printer.

During the 6 and 10 P.M. broadcasts, I warned the viewers about the possibility of severe weather. While there was no way of knowing what would happen, I was anxious. Something just felt wrong.

I drove home that night with the window rolled down, listening to college basketball scores on the radio. My wife and I cooked out on the back porch, but the wind was so strong we had to move the grill just to get it fired up. Despite the wind, we sat out in our summer shorts, looking at the evening sky and enjoying the warmth.

I mentioned I was concerned about the possibility of bad storms. My wife said there was no sense worrying about what you can't control. But despite the perfectly grilled hamburg-

Storms hit the Midwest

Temperatures hit record highs again Tuesday, reaching the 70s and 80s in the East, and thunderstorms rattled the southern Plains and Illinois. Snow fell in Colorado and Wyoming.

A tornado touched down near Jetmore, Kan., during the morning but no injuries or major damage were reported. Tornado watches were posted for parts of Kansas, Nebraska, Iowa, Missouri, Oklahoma and Texas.

Hail as big as golf balls fell in Illinois at Mazon and near Elvira, the National Weather Service said. Three-quarter-inch diameter hail fell at Pampa, Texas.

Winds gusted to 65 mph at Gage, Okla., and 40- to 50-mph wind was reported in central and north-central Kansas.

COURTESY OF QUAD-CITY TIMES

ers, when we finally shut out the lights and the balmy air rustled the shades in our bedroom, I was uneasy.

The next day, March 13, dawned bright, windy, and warm. There was a fresh, almost sweet smell to the air that comes with the first muggy days of spring. I recognized the scent immediately. It indicated the type of day thunderstorms thrived in.

I dressed in a dark suit and a red silk tie, knowing that if the storms popped, I would have extra air time and I had to appear professional and authoritative.

That afternoon, when the station's chief engineer, John Hegeman, and I walked into the Civil Defense office to meet Bud Whitfield, I noticed thunderheads in the southwest sky. Whitfield and I discussed warning procedures and ways to improve communications, and in no time I was on my way to work.

Cruising back up the Brady Street hill on the way back to the station, I noticed that the northwest horizon was filled with anvil-shaped thunderheads. Lightning flickered from the base of the clouds. I nudged John and told him I thought things were going to get active.

With a loud, pinging sound, the first warning of the day tripped the weather alarm at 1:30 P.M., just minutes after I strolled into the weather center. I had programmed the alarm to alert me if the weather service issued any sort of warning for specific counties. The storm cluster I had seen was spitting hail. On the police scanner, people were describing hail the size of golfballs and winds approaching 60 miles per hour. One of the most alarming reports described a hailstorm so intense that it covered fields and a section of Highway 61 between Eldridge and Park View with a mantle of ice.

Delores Rathjen, in the Park View Dairy Queen, talked about her experience with the hailstorm. "It sounded like someone was shooting at me," she said. "I saved some of the pieces of hail in the freezer. I've lived here all my life and never saw anything like this."

About this time, low pressure streamed into western Iowa on its journey north to the Great Lakes. Stretched across its center was a cold front that split Iowa in two before it continued south into Texas. On the eastern edge of the front, thunderstorms exploded. Tornado watch boxes peppered the Midwest. Watch boxes are issued by the Storm Prediction Center for specific geographical areas threatened by severe weather. Each has a specific starting and ending time. Tornadoes had already touched down in several states. And it was still early in the day.

The temperature on the digital thermometer in our weather center registered a balmy 75 degrees—a record high. After a brief rain shower, the sun had re-emerged and the added humidity was noticeable. The atmosphere was charged and extremely unstable. Like a gun, it was loaded, cocked, and ready to blow. All it needed was an advancing cold front.

After the initial wave of storms skirted off to the north, I had time to take stock of the situation. I met with director Mark Koster and the newsroom staff and told them the severe

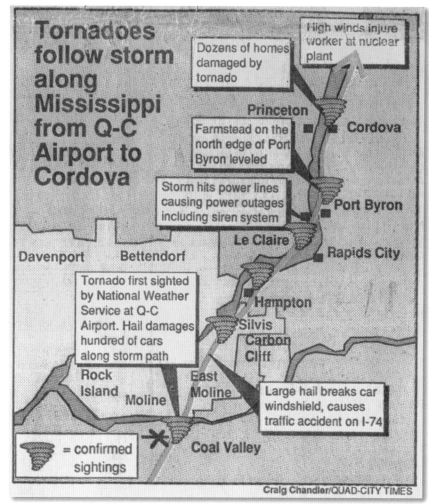

Tornadoes follow storm along Mississippi from Q-C Airport to Cordova

High winds injure worker at nuclear plant

Dozens of homes damaged by tornado

Farmstead on the north edge of Port Byron leveled

Storm hits power lines causing power outages including siren system

Tornado first sighted by National Weather Service at Q-C Airport. Hail damages hundred of cars along storm path

Large hail breaks car windshield, causes traffic accident on I-74

Princeton · Cordova · Port Byron · Le Claire · Rapids City · Davenport · Bettendorf · Hampton · Silvis · Carbon Cliff · Rock Island · Moline · East Moline · Coal Valley

= confirmed sightings

Craig Chandler/QUAD-CITY TIMES

weather threat had not passed and that more storms and more weather cut-ins—or interruptions in regular programming— were likely. Even so, the radar was dark and deceptively clear. It was 4 P.M. and I was waiting.

The cold front was getting close, and I was beginning to wonder if these storms were going to pop. It wouldn't be the first time. I took my forecast to a chyron operator in A-control, the room where the director, producer, audio, and chyron operator work when the broadcasters are on the air.

The 1990 tornado as it hit Port Byron, Illinois, at 5 P.M. COURTESY OF RIVER VALLEY LIBRARY, PORT BYRON, ILLINOIS

When I returned to the weather center, my eyes instinctively turned to the radar where two bright spots—thunderstorms—glowed just to the southwest of the Quad Cities. The waiting game was over.

As I watched the storms develop on the radar screen, it soon became apparent that the cell in Mercer County was one to reckon with. With each sweep of the radar beam, its intensity was increasing so much that the supercell was bound to become severe. The storm was heading northeast toward Rock Island County. I calculated that it would reach the southern part of the Quad Cities in less than 15 minutes.

Over the next five minutes, the storm evolved into a massive thunderhead. Scanning its structure by way of radar, I noted a hail shaft rapidly forming. I checked for rotation. Nothing obvious showed up that indicated that the storm was twisting.

However, the isolated nature of the mushrooming storm concerned me. As an apparent supercell, it was not competing with other storms for the heat, moisture, and instability that had led to its development. It was the type of storm that was capable of doing exceptional things.

The phone rang. An excited viewer, calling from out near the Moline airport, said golfball-sized hail had been falling for several minutes. I was skeptical until he stuck the phone out the door and I could hear the hail pounding on the porch roof.

That was all I needed. The storm was showing its hand. It was time to get on the air.

As I told my director that it was time to go on the air with a warning, a station employee walked in from the parking lot and said he thought he saw a wall cloud to the south. Seconds later, a newsroom producer raced in saying the scanners on the

Weather Folklore

When the leaves of trees turn over, it foretells wind or severe weather.

Illinois side were reporting hail, and funnel clouds had been sighted near the Moline airport.

Then I saw pictures on a computer monitor from our live skycam—a camcorder mounted high on the station's tower. I was unprepared for what I saw. Less than 4 miles away, a classic wall cloud rotated counter-clockwise at the base of a tremendous thunderstorm. Big, ominous, and remarkably surreal, it was a captivating cloud. And it was headed right for the heart of Moline.

Grabbing a microphone, I raced for the set, forgetting the suit jacket I had so carefully picked out. In the meantime, the printer in the weather center spat out the information I already knew—a severe thunderstorm with a possible tornado moving into the Quad Cities metropolitan area. The National Weather Service issued an official tornado warning for Rock Island County in Illinois and eastern Scott County in Iowa. It was 4:25 P.M. Seconds later, I was on the air with live pictures of a developing tornado.

The newsroom was a flurry of activity as the gravity of the situation increasingly became clear. Reporters and photographers grabbed gear and ran out the door. The atmosphere of the newsroom became as charged as scanners chattered, phones rang, and people shouted. As information poured in, I had to get it on the air, and fast.

As the skycam rolled, I alternated between showing pictures of the Doppler radar and pictures of the actual storm. It was a unique experience to explain the structure of the storm on radar and then show the results with live pictures from the skycam. It was like teaching Tornado Development 101 to a class of thousands.

Every few seconds, someone passed on some new information. One report of ping-pong sized hail was handed over on a napkin smeared with ketchup. Another report scrawled on a piece of paper was so illegible that I could only make out bits and pieces of the message.

As the chaos continued, the storm surged up the Mississippi River, which was instantly flashed across the

The March 13, 1990 tornado as it approaches the home of Scott Verbeckmoes. PHOTO BY SCOTT VERBECKMOES

53

screens of thousands of television sets. Hail the size of baseballs was driving into the ground across parts of Moline and Rock Island, Illinois, breaking windows and denting the bodies of thousands of cars, stripping leaves from trees and severely damaging the roofs and sidings of a number of homes. As the wall cloud pressed on, we received regular reports of funnel clouds.

I had been on the air for 15 minutes. As I continued with my coverage, our news director decided the situation was so serious and fast-breaking that we should include anchors and reporters. Since I was on the air nonstop, this decision was not apparent to me until Paula Sands showed up on the

Bob Hunt and his wife waited out—and survived—the March 13, 1990 tornado in his home's basement.
COURTESY OF RIVER VALLEY LIBRARY, PORT BYRON, ILLINOIS

Highway 84 in their patrol car. In the newsroom, their observations crackled across the scanners. The urgent words were anything but good: the small communities of Port Byron and Le Claire stood directly in the twister's path. As the two towns braced for the arrival of the storm, golfball-sized hail pelted the adjoining Mississippi River, producing plumes of water 3 feet high as they peppered the river.

On the set, I couldn't show the storm in any kind of detail because the tornado had now moved too far away for the skycam to capture it in detail. I was left with the Doppler radar as my primary tracking device. While the colored pixels of the radar images may not have been as captivating as pictures of a live tornado, they were still compelling. As I told the viewers how the storm's strong rotation was wrapping the winds into a tight circular pattern known as a hook echo—the signature of a classic tornado—I pointed at the swirl of color on my weather map that lay just above the towns of Port Byron and Le Claire, Iowa. There was no need for calculation. The storm was seconds away from tearing into these towns. I warned the viewers that if they were in the path of this storm, especially in Port Byron or Le Claire, they had to seek shelter immediately!

About this time, Scott Verbeckmoes of Port Byron told me later, he stood up from his chair in front of the television where he had been watching my coverage of the tornado, and went to the window. "I had seen the television skycam pictures from Iowa, and 5 minutes later, the same silhouette was outside," Verbecmoes said. "I sent my wife, Iona, down to the basement and went outside. It came closer and lifted at the river—with its tail up. Then it came down our side and formed on the Illinois shore. It kept growing and came within 200 yards from my house. The tail was at treetop level. I took off for the house."

Amazingly enough, the tornado ran through the small gap between the towns of Le Claire and Port Byron. As it churned up the Mississippi, residents on both sides of the river watched the tornado as it sucked river water into its

set of her talk show with a microphone and a handful of storm reports. I moved to the set, where I was glad to give someone else a chance to talk while I took a breather.

By now, the storm had moved between Bettendorf and the north edge of Moline, and we were losing our ability to show it live. At this point, the first confirmed tornado touchdown reached the news desk. Moments later, Al Carter, a photographer chasing the storm, radioed in from the I-80 Bridge near Le Claire, Iowa. As the remaining newsroom listened in, Carter calmly reported, "Tally ho, we have a touchdown!" As his camera began to capture images, the tornado skipped northeast along a path that paralleled the Mississippi.

From the outskirts of Hampton, state troopers Robert Elliot and Dennis Demaught radioed reports to their dispatchers as they followed the tornado along Illinois

belly. The tornado did not make landfall until the north end of Port Byron. In its path was the 128-year-old house of Bob Hunt.

Hunt gave this account in the March 14, 1990 *Quad-City Times* as he gripped a mud-stained bible. "I was just down at the barn when my wife yelled, 'Come quick, there's a tornado warning,'" Hunt said. "I ran to the house and grabbed a beer. Then I looked out the window and saw it coming. The wife yelled at me to get in the basement."

Within seconds after reaching the basement, the tornado ripped into Hunt's home and reduced it to a pile of rubble. "It didn't shake or make a noise; it just went woof." He and his wife escaped, shaken but unharmed. "God was just looking out for us. It was God and a smart wife that got me through this."

Next on the hit list was Cordova.

The station's Doppler radar now confirmed a new development. The storm was veering from its northeast course and was headed straight north toward Cordova, Albany and Fulton. And the hook was more defined than ever. With fresh word of damage and more funnel cloud sightings, I urgently pled with viewers to take cover.

Moving up the east side of Iowa Highway 84, the storm sliced a path through a stand of oak trees, tearing out power lines along the way. On the outskirts of Cordova, the tornado spun through a cemetery, throwing tombstones right and left before taking aim at the north side of town.

As sirens blared, Jeff Staken saw the black cloud approaching. "It looked like it was coming straight for me," Staken said later in the *Quad-City Times*, "As I took cover in the basement, I could feel the house shaking. I came back up to see [the tornado] sweeping back and forth like a broom, kicking up shingles and debris."

Nearby, 14- and 15-year-old Vanessa and Megan McIntyre rode out the storm in their house. As they talked to the *Quad-City Times* reporter, Rod Thomson, their eyes were red from crying. "We saw the clouds and the twisting

Bob Hunt and the tattered family Bible both survived the tornado. COURTESY OF *QUAD-CITY TIMES*

55

Toppled gravestones in the Cordova cemetery. COURTESY OF QUAD-CITY TIMES

and ran downstairs," Megan said. Vanessa said she heard a big explosion, "just a boom," and then "all this stuff, insulation, came falling down around us." "I about freaked," Megan said. They escaped unharmed but their house lost its roof.

In minutes, 35 homes were damaged or destroyed in the area surrounding Cordova. The homes along North River Road took the biggest hit—23 were damaged or destroyed.

Before lifting from the ground, the twister headed toward one more target—the Cordova nuclear power plant just north of town. With just 25 minutes to prepare, nearly 300 workers huddled in the main plant to await the tornado's arrival.

The tornado roared into the parking lot, twisting and turning, at 5:05 P.M. It hugged the shoreline of the Mississippi River as it battered the perimeter of the facility, striking outbuildings and mobile construction units and inflicting thousands of dollars in damage. Then the storm streaked narrowly by. Inside the plant, the workers let out a collective sigh of relief. Miraculously, the reactor, turbines, and the electric lines were not damaged. Outside in the fading sunlight, the tornado spun back into the clouds and was gone forever.

As I examined the radar, a change in the storm was clearly evident. The rotation had diminished and the amount of red and yellow returns had diminished. The storm was weakening. and the worst was over. Around town, damage was widespread, and power was out. Sirens filled the air as emergency crews sped to assist those in need.

From here on out, the story belonged to the anchors and reporters. It was time for the newsroom to turn its attention to those in the path of the storm. They needed to determine if there were any fatalities and, if so, how many. They needed to figure out how many people were hurt and assess the damage on the buildings.

The barn in the foreground was completely destroyed by the impending tornado. Note the golf ball-sized hail in the foreground. PHOTO BY JOHN SAMPLE

THE DAILY ... ATCH

MOLINE, ILLINOIS   35¢

WEDNESDAY, MARCH 14, 1990

112th YEAR — No. 224

Tornado warning saves lives

Upper county takes shelter as twisters dance along river; 35 homes hit

My job was done. I took a deep breath and chugged a can of Diet Mountain Dew.

Back in the weather office, the phones were ringing non-stop and paper littered the floor. I looked up at the clock and was astonished that it was 5:30 P.M. I was shocked. I had been on the air for over an hour straight. I was so focused I had completely lost track of time.

That night, I learned that although there were injuries, there were no fatalities from the storm. Thirty-eight homes were damaged or destroyed, and damage was estimated in the millions, but not a single person had been killed. Luck and ample warning gave people the chance to find shelter. I was proud to have been a part of that.

As the evening wore on, home videos of hail and tornadoes were delivered to the station. Never had a tornado been documented by so many people in the Quad Cities. On the 10:00 P.M. news, we were able to use home videos of hail and tornadoes along with the work of our own photographers to show a number of images of the storm.

We showed several hailstones the size of baseballs that were delivered to us by viewers in Moline. We kept them in the freezer until just before news time. One jagged ball of ice measured 3 inches in diameter.

Several days later, I learned more details about the tor-nado. The National Weather Service determined that one or more tornadoes skipped and looped in an erratic, north-easterly direction from 1 mile southeast of Riverdale, Iowa, to 3 miles north of Cordova, Illinois. At its peak, the torna-do was 250 yards wide and, near Cordova, exhibited multi-ple vortices that destroyed or severely damaged 12 homes. Multiple vortices are essentially mini-tornadoes within the main tornado. They are common in stronger twisters and explain why damage can be more severe in one part of a tor-nado than in another. The tornado, with winds of 158 to 206 miles per hour, was classified as a severe F-3 tornado on the Fujita Scale. For the record, only 25 percent of all tor-nadoes ever reach F-3 status.

But when I got home that night, the full impact of the day finally hit me. I was exhausted. At 1:00 A.M., I flopped into bed. When I closed my eyes, I saw the silhouette of a tornado spinning through my mind. It was then that I real-ized what an exceptional moment I had experienced. I had taken on an F-3 tornado and covered it, live, for thousands of stunned viewers. For whatever reason, I was grateful to be the one chosen for the task. And, despite the twister's best efforts, it hadn't claimed a single victim.

In the darkness, I managed a smile before drifting off to sleep.

Thunder, Lightning & Hail

There are few things I enjoy watching more than a rip-roaring electrical storm. Yet, while many of us consider lightning great entertainment, early cultures were not so enamored with nature's light show. The Greeks, for example, used mythology to explain lightning and ease the fears of those who witnessed it. They believed that Zeus, the king of all gods, threw lightning down from the heavens to show his anger at the people below.

Even though the world eventually stopped thinking of lightning as a weapon, it was still feared and respected until the 1700s, when scientists like Benjamin Franklin made some illuminating discoveries. By flying a kite during a thunderstorm, Franklin determined that lightning was an electrical current when he observed sparks flying from the key he had tied to the kite string. He was lucky to have survived, but this discovery of the electrical properties in thunder was crucial in furthering our understanding of electricity.

Lightning is one of the most beautiful displays in nature, but it is also one of the most deadly and unpredictable phenomena known to man. A lightning bolt is hotter than the surface of the sun (54,000 degrees) and, as it pierces the atmosphere, it sends shockwaves in all directions.

This shockwave is what we know as thunder. Thunder is caused when the air rapidly expands because of the intense heat of the stroke. This expansion of air that creates what is

FRANKLIN'S EXPERIMENT WITH THE KITE.

LEFT: *An 1877 rendering of Benjamin Franklin's famous kite experiment, when he set out to prove that lightning was electricity.*
COURTESY OF NOAA/ HISTORIC NWS COLLECTION

FACING PAGE: *An electrifying moment is captured.*
PHOTO © JIM REED; WWW.JIMREEDPHOTO.COM

What to Do During a Thunderstorm

During a thunderstorm, lightning can strike up to 10 miles away from the rain area. Thunder cannot be heard from storms more than 10 miles away, so it may be difficult to tell that a storm is coming your way. The best rule of thumb is that if you can hear thunder, you are within striking distance and you should seek shelter immediately.

The 30-30 Rule

Where there is nothing to obstruct your view of a thunderstorm, you can use what I call the 30-30 Rule. When you see lightning, count the seconds until you can hear thunder. If that time is 30 seconds or less, the thunderstorm is within 6 miles and is dangerous. Seek shelter.

Wait at least 30 minutes after the last clap of thunder before leaving shelter. The threat of lightning continues for a much longer period than most people realize.Basically, as long as you see lightning and hear thunder, you are close enough to a storm to be in danger. Don't be fooled by sunshine or blue sky!

Multiple cloud-to-cloud and cloud-to-ground lightning strokes.
COURTESY OF NOAA/PHOTO LIBRARY

Don't Go Swimming During a Thunderstorm

In the United States, most lightning deaths and injuries happen during the summer months when lightning storms are active and people are involved in outdoor activities. When thunderstorms approach, if you are outside bicycling, boating, camping, fishing, golfing, hiking, jogging, swimming, walking, or working, seek shelter. Coaches, umpires, referees, or camp counselors should halt activities to make sure that the participants have time to get to a safe place as soon as the storm is detected.

Inside homes, you should also avoid activities that might endanger your life should lightning strike. People should stay away from windows and doors and avoid contact with anything that conducts electricity, such as phones or computers.

You might also want to take precautions to protect property, such as electronic equipment. Some think surge protectors keep electronic gadgets—such as computers and televisions—safe during electrical storms. Surge protectors, however, will not protect your equipment from lightning strikes. You need a lightning arrester to protect your equipment. This device uses a gas-filled gap that becomes ionized and conducts energy when a higher surge of electricity, such as a lightning strike, is detected. As the lightning enters the electrical line, the gas cap safely conducts the current to the ground, where it dissipates.

Helping a Lightning Strike Victim

If a person is struck by lightning, immediate medical care may be needed. People who have been struck by lightning can experience cardiac arrest and irregularities, burns, and nerve damage. With proper treatment, including CPR, if necessary, most victims survive a lightning strike. But the long-term effects on their lives and the lives of family members can be devastating.

known in physics as a compression wave. As it hurtles through the sky, it quickly manifests itself as a sound that rattles the windows and sends the cat under the bed.

Since sound travels roughly 1 mile every 4.5 miles per second and light blazes along at 186,000 miles per second, during a thunderstorm, we see the flash of lightning before we hear the thunder. You can estimate the distance of the thunderstorm producing the lightning by counting the number of seconds between the lightning and the thunder.

If there are 4.5 seconds between the lightning and the thunderclap, the storm is 1 mile away; 9 seconds' difference means the storm is 2 miles away. You will get the full symphony of light and sound simultaneously when the strike is directly overhead.

Thunderstorms also produce static. If you don't see lightning or hear thunder, but you suspect a thunderstorm is in the area, one way to find out without radar is to turn on an AM radio. If you dial in a frequency on the left end of

A dramatic cloud-to-ground lightning display. PHOTO © JIM REED; WWW.JIMREEDPHOTO.COM

Weather Folklore

If the bull leads the cows to pasture, expect rain; if the cows lead the bull, the weather will be uncertain.

the dial where there is no station and hear static, it is most likely caused by a storm. You will know that a thunderstorm is on its way if the static gets louder and more frequent.

Another interesting fact about lightning is that it combines with atmospheric gasses to produce what is known as oxides of nitrogen and ozone. These natural fertilizers keep things green and growing. Sometimes,

just before a thunderstorm, you can smell the sweet, damp smell of the oxides as the rain approaches. This type of fertilizer smells a lot better than the fertilizer our farmers spread over their fields!

On a stifling night during the dog days of summer, you may have seen the flash of distant lightning and been told it was heat lightning. There is no such thing. What you are actually witnessing is a thunderstorm that is up to 100 miles away. Thunder cannot be heard from storms more than 10 miles away, which has led some to believe these thunder-free storms are the result of heat. This is incorrect. Heat provides

A dazzling cloud-to-cloud discharge of lightning. PHOTO © JIM REED; WWW.JIMREEDPHOTO.COM

energy that will help create a thunderstorm, but heat alone will not cause one.

Lightning generally occurs during warm weather. If a storm is strong enough, however, lightning can strike in winter even if temperatures are well below freezing. During our worst winter storms, "thunder snow"—snow accompanied by lightning and thunder—occurs.

These convective snow bursts can produce staggering amounts of snow—ranging from 2 to 3 inches an hour. A convective snow means there is a vertical structure large enough in a storm for convection, meaning lightning and thunder. Most winter precipitation forms in clouds that are flat or stratiform in nature. They rarely extend higher than 15,000 feet. During powerful winter storms, the dynamics that drive them are so strong they can force the clouds higher than 30,000 feet where conditions for thunderstorms can be achieved.

In 1994, a national lightning network was created that is operated by

Weather Folklore

**When the dew is on the grass,
Rain will never come to pass.
When the grass is
dry at morning light,
Look for rain before the night.**

Multiple streamers light up the night sky. PHOTO © JIM REED; WWW.JIMREEDPHOTO.COM

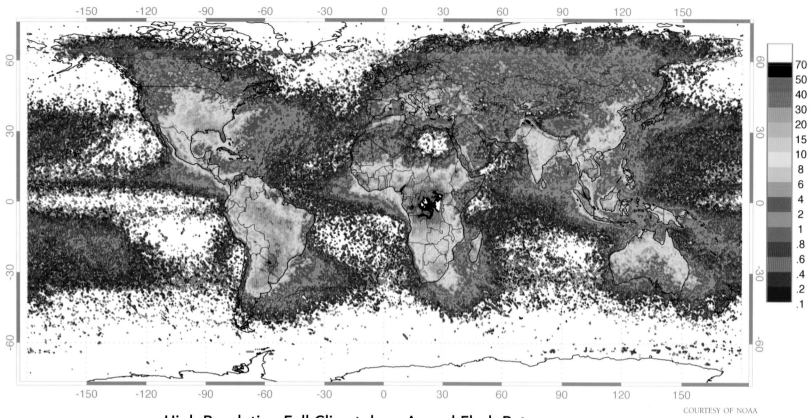

COURTESY OF NOAA

High Resolution Full Climatology Annual Flash Rate

Global Distribution of Lightning April 1995-February 2003

Global Atmospherics in order to determine just how much lightning actually occurs in the United States. With data from a network of antennae around the country, meteorologists have discovered that there are an average of 20,000,000 cloud-to-ground flashes a year. They have also discovered that lightning strikes 30 million ground points—places in the ground where the lightning bolts hit—every year.

Lightning can be deadly. Every year in the United States, approximately 300 people are struck by lightning and approximately 70 will die from the strike. Lightning is every bit as deadly as it is beautiful, and it needs to be taken seriously. It is very important that you respect its power and know what to do when it makes its presence known.

When lightning kills, the power and heat of the bolt (often as hot as 25,000 degrees Celsius) can literally blow the shoes off a person. As the electricity travels through the body, it can sear, mangle, and vaporize tissue. A victim may appear normal on the outside but have severe internal injuries. Even if a person is lucky enough to avoid internal injury, the shock of the charge can easily stop the heart.

— Pennies, Golfballs, and Baseballs: The Many Faces of Hail —

Hailstones are the product of thunderstorms. They begin life as tiny dust or ice particles, similar to the nucleus of an atom. In the updrafts and downdrafts within a thunderstorm, these particles circulate into areas of super-cooled droplets of water—water that is unfrozen but in an area where temperatures are below 32 degrees. The droplets freeze when they attach themselves to the particles. The particles grow larger as they move up and down within the thunderstorm and eventually become so heavy that they fall to the ground as hail.

The Quad Cities area averages less than one day per year with hail at least 2 inches in diameter. Damaging hail of that type is most likely found over the high plains extending from North Dakota to Texas. Gorilla hail, up to 4 or 5 inches in diameter, is usually confined to Oklahoma and Texas as the bull's-eye on page 66 shows.

Each time a hailstone circulates in the core of a storm, it gathers a new layer of ice. By cutting a hailstone in half, you can clearly see these layers or rings of ice. If you count the rings in the hailstone's core, you can determine how many times the stone circulated within the storm before it fell to the ground. A hailstone's size is also an indication of the strength of a thunderstorm's updrafts. The stronger the updrafts, the larger the hailstone will be.

The analogies people use to describe the size of a hailstone always amuse me because they say so much about the cultures of the people who are making the comparison. Since hailstones are generally round, they are, of course, compared to other round objects. Around the country, hailstones are regularly compared to baseballs, golfballs, grapefruit, hens' eggs, marbles, peas, ping-pong balls, and softballs. In some circles, money is used as a measuring stick—hailstones are compared to nickels, dimes, quarters, and half

A handful of gorilla hail. PHOTO © JIM REED; WWW.JIMREEDPHOTO.COM

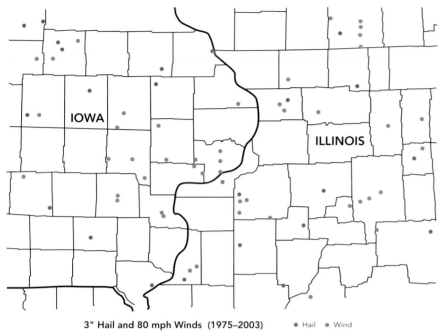

3" Hail and 80 mph Winds (1975–2003)   ● Hail   ● Wind

Hail from 2.76 to 4.50 inches (green dots) and winds of up to 80 mph (red dots) from 1975 to 2003.
COURTESY OF NATIONAL WEATHER SERVICE, DAVENPORT, IOWA

66

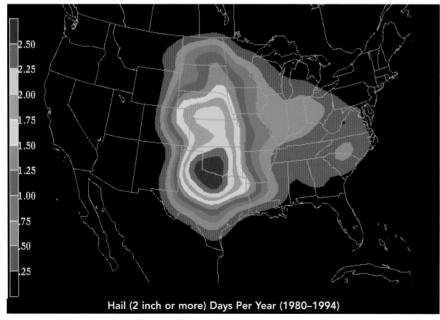

Hail (2 inch or more) Days Per Year (1980–1994)

COURTESY OF NOAA/NATIONAL SEVERE STORMS LABORATORY

dollars. In India, giant hailstones have even been compared to "cricket balls."

Whatever comparisons are used, a hailstone that is three-quarters of an inch in diameter (about the size of marble or penny) is likely to cause damage when it falls. The National Weather Service considers a storm that contains hail this size as severe, and it will issue a storm warning.

On May 15, 1968, I was in the fifth grade in my home-town of Coralville, Iowa. As I munched my way through my favorite breakfast of toast with brown sugar and cinnamon, I had no idea that that the day was going to be one of those days in weather history when the weather gods decided to throw everything nasty they had at eastern Iowa. During one of the worst modern-day tornado outbreaks in Iowa history, severe thunderstorms ravaged the countryside from morning until late evening, spitting out hail and spinning up twisters that left random trails of destruction.

The worst of the batch blasted Charles City at 4:47 P.M., when a wedge-shaped, F-5 tornado skipped through town and killed 13 people. Ten minutes later, a second twister struck Oelwein and Maynard, Iowa, and left five more people dead. The violence of the storms stunned Iowans.

Tornadoes, fortunately, skipped Coralville, Iowa. What did happen, however, was a once-in-a-lifetime hailstorm.

My first inkling of trouble came when, mid-morning, my teacher turned on all the overhead lights. Outside, the streetlights had come on as the sky turned an ominous dark shade of green. Thunder had begun to rumble when there was a knock on the classroom door. It was the principal, who told us that we were to crouch next to the brick wall on the eastern edge of the classroom and cover our heads.

Huddled with my classmates, I could not see what was going on outside, but I could hear enough to know it was not good. Stillness had enveloped the room as we looked at one another, then away. The silence continued until, suddenly, it sounded like rocks were falling on the roof. At first there were just a few thumps, then a steady din, and finally a thundering roar as hail descended. As we waited out the storm, I remember the sounds of whimpering kids.

In minutes, it was over. We rushed to the windows. There, coating our playground from the ball field to the parking lot, was hail the size of golfballs. Even though it was spring, it looked like Christmas. When we were let out for recess several hours later, hundreds of chunks still remained. My pals and I had a fine time throwing hailstones at one another until the teachers called for a end to the fight.

When Dad pulled into the driveway that evening, his car didn't look anything like the one he left in. Dents and pockmarks covered the body of the car he so meticulously cared for. The storms had hit hard in Iowa City, where he worked, and it had given our Chevy Caprice a pretty good pounding.

Later, other townspeople told us about hailstones in Iowa City measuring 11 inches in circumference—almost the size of softballs. Many ice chunks were so big they remained frozen for much of the afternoon, despite temperatures of up to 80 degrees.

Since that time, I have seen many a storm come and go, but I clearly remember the fear that hailstorm etched across the faces of my teachers and classmates. The sound of those ice balls, traveling at speeds of more than 100 miles per hour, pounding on the school roof, terrified us with its fury.

Due in large part to the National Weather Service's advanced warning system, few people in this country die from hail-related injuries. In other parts of the world, however, the news is not so good. As recently as 1986, a hailstorm in the Sichuan province of China injured 9,000

A tornado approaches Charles City, Iowa.
PHOTO COURTESY OF FLOYD COUNTY HISTORICAL SOCIETY

Killer Hail

According to official records, only two deaths in U.S. history have been attributed to hail. One was a Texas farmer in 1930, the other a three-month-old baby in 1979. In other parts of the world, especially India, hail has proved far more deadly. In 1888, hailstones the size of oranges killed 246 near Moradabad, India. Many victims were reportedly pounded by hailstones while others died from exposure when they were buried in icy drifts 3 to 5 feet deep. At least 1,600 animals died in the storm.

Killer Lightning

Of all the world's weather phenomena, lightning is one of the most deadly. Every minute of every day, 1,800 thunderstorms are creating lightning somewhere on the Earth. Each year in the United States, about 70 people are killed by lightning and another 300 are injured. Eighty-four percent of the victims are male and 73 percent of the fatalities occur during June, July, and August. The majority of accidents occur on weekends around 4:00 P.M.

67

This nearly 8-inch hailstone holds the record as the largest in U.S. history.
PHOTO BY QUILLA ULMER/JIM REED PHOTOGRAPHY

Hail 2 to 3 inches in size piles up on streets and grass during a severe thunderstorm.
COURTESY OF NOAA/NWS PHOTO LIBRARY

68

What To Do During A Hailstorm

To be safe in a hailstorm, stay inside and stay away from windows and any glass. If you are in a car, pull over until the storm passes. Also, check the size of the hail. If the hailstones are as large as golfballs or bigger, that often foretells the possibility of a tornado. Check all the latest watches and warnings from the National Weather Service.

people. Another 100 died from hail-related injuries, many of which were skull fractures.

On the other hand, hail is responsible for nearly one billion dollars' worth of damage to crops and property each year in the United States. The costliest U.S. hailstorm battered Denver, Colorado, on June 11, 1990, when golfball- to baseball-sized hail fell over a wide area of the city and did $625 million in damage. Denver insurance agents wrote thousands of checks and worked many hours of overtime to compensate for the effects of that storm.

For the most part, the hail in the Midwest ranges from pea- to golfball-sized. On rare days, however, supercell thunderstorms can become powerful enough to produce the updrafts necessary to cause "gorilla hail"—hailstones the size of softballs or grapefruit. Hail this large can actually produce holes in roofs and strip trees of their leaves, bark, and in some cases, branches.

An example of gorilla hail would be the hailstones that fell in Coffeyville, Kansas, on September 3, 1970. The hailstones measured 17.5 inches in circumference and were 5.5 inches in diameter and weighed 26 ounces or 1.6 pounds. For more than three decades, this was the largest hailstone in U.S. history.

These days, however, the distinction for the largest hailstone belong to Aurora, Nebraska. Following a thun-

Large Hail Reports by Month
1955-1993 / WFO DVN Modernized CWA

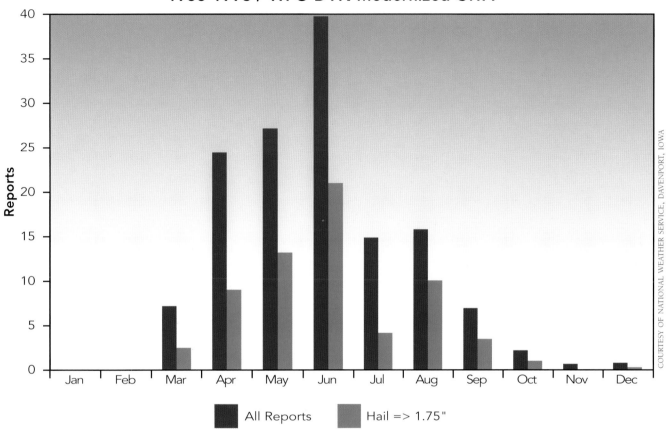

COURTESY OF NATIONAL WEATHER SERVICE, DAVENPORT, IOWA

All Reports   Hail => 1.75"

derstorm on June 22, 2003, a hailstone was recovered measuring 7 inches in diameter and 18.75 inches in circumference. When it fell, the hailstone created a crater on the ground that was 3 inches deep and 1 foot wide! This hailstone is now enshrined in the National Center for Atmospheric Research in Boulder, Colorado, where it will be preserved indefinitely.

The largest hailstone ever measured in Iowa fell to the ground in Dubuque, Iowa, during a storm on June 16, 1882. This 1-pound, 12-ounce hailstone with a circumference of 17 inches and a diameter of roughly 5.4 inches was slightly larger than a softball.

The sky turns green in a storm when there is hail.

A statute holds that a river has a right to overwhelm its banks and inundate its floodplain. Well, that's interesting because it's not a right that we assign to the river. The river has earned it through centuries of deluging and shaping the floodplain and the floodplain has a right to its rampaging river.
—DAN KEMMIS, HARPER'S, FEBRUARY 1991

Floods

The arrival of spring is one of the great rewards of living in Iowa. For me, spring is more than a date on the calendar. Spring finally arrives on a sun-kissed day in May, not during the chilly month of March or windswept April. It's the riot of color that suddenly arrives: the green grass, the red tulips, and yellow daffodils. It is lying in bed with the windows open and hearing the crickets chirp. Spring is the exhilaration I feel when winter is finally, unmistakably, gone.

The tranquility of those gorgeous days, however, belies the fact that our weather is at its most violent in the spring. The warmth that feels so good brings with it moisture—the atmospheric fuel that combines with the heat and can generate towering thunderstorms and torrential rains. If conditions are right, these rainstorms can release tremendous amounts of water over a short period of time. These are known as flash floods. Flash floods develop quickly, come

RIGHT: *The Duck Creek Flood of 1990.* COURTESY OF *QUAD-CITY TIMES*
FACING PAGE: *Mammatus clouds, named for their shape.*
PHOTO BY MIKE HOLLINGSHEAD

Raining Cats and Dogs

Ever wondered where the phrase "it's raining cats and dogs" originated? Some say the explanation dates back to medieval times when people lived in houses with thatched roofs. Their pets, in this case cats and dogs, lived outside the huts. At night, the cats and dogs would climb onto the roof to enjoy the warmth rising from within the house. When a heavy rain occurred, the straw became slippery and occasionally a cat or dog fell through the roof. Hence, another saying: "Heads up!"

without warning, and can be catastrophic, destroying lives, homes, and property.

While flash floods do not get the attention and publicity that tornadoes get, flash floods are a formidable foe. In the U.S., flash floods kill more people each year than hurricanes, tornadoes, windstorms, and lightning, making them the nation's number one weather-related killer. They cause an estimated one billion dollars' worth of damage to homes and businesses each year. In terms of sheer calamity, flash floods are under-appreciated.

Flash floods have increased in recent years because of urban sprawl. As our cities expand, the land around small rivers and streams that once absorbed rainfall is now covered with concrete. Instead of soaking into the earth and down into the water table, the rain runs off the non-porous surface. Nature no longer has a way of disposing of excess rain. Instead, the additional runoff expands the flood plains. A rainstorm that, in years past, may have created minor problems can become a significant flood.

A flash flood, in its simplest form, is the result of too much rain over a short period of time, usually two to four

Beautiful but dangerous, a thunderhead explodes into the evening sky.

hours. The Duck Creek Flood of 1990 was a classic example of a summer flash flood that was the result of what meteorologists call a mesoscale convective complex. Common to the Great Plains and the Midwest, a mesoscale convective complex (or MCC) is a cluster of storms over 50 to 100 miles wide that happens in June, July, and August, when the steering currents of the jet stream are weak. This makes for slow movement of the MCC and some incredible rainfall totals that can exceed 10 inches in a few hours' time.

During late spring, another factor that contributes significantly to heavy rain and flooding is the presence of the Great Lakes—Lake Michigan in particular. When it is still

A massive thunderhead billows into the sunset. PHOTO BY MIKE HOLLINGSHEAD

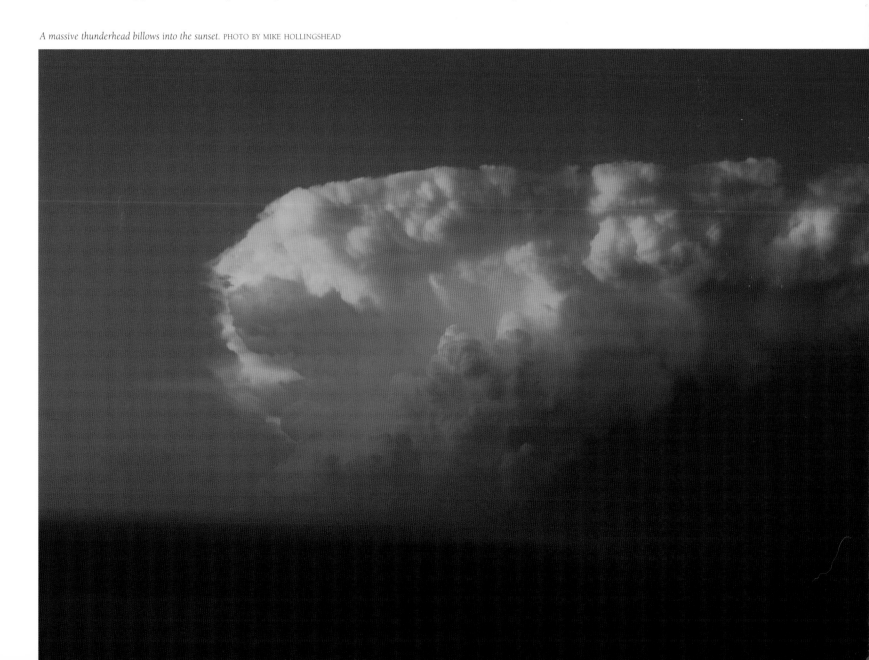

Part Time Rainbows

Most of us enjoy the beauty of a brilliant rainbow. However, have you ever noticed that you never witness them at noon? The altitude of the sun above the horizon determines the amount of a rainbow's arc that you can see. Once the sun climbs over 51 degrees above the horizon (about midday), the sun's angle is too direct to create the magical effect. As the sun sets later in the day, you will get another chance to see a rainbow, weather permitting. And, don't forget, your back must be to the sun to see a rainbow!

What To Do During a Flood

Flash floods, big or small, have one thing in common; they have a way of enticing people to drive through what appears to be shallow water. This is a mistake that has proven fatal over and over again. Statistics show that nearly 60 percent of all the people who die in floods do so when their vehicle is swept away by moving water. As little as six inches of swift moving water can float a small car. Water moving at 4 miles per hour, a speed that is equivalent to a brisk walk, exerts a force of 66 pounds on every square foot it encounters. Increase the water speed to just 8 miles per hour and the force explodes to about 260 pounds per square foot. That will carry away a small truck when the water merely reaches the top of the wheels.

The bottom line is to never drive or wade into floodwaters. Just a few inches of water over a road can wash away pavement or the bottoms of bridges. What appears to be a shallow layer of water may actually be a washout several feet deep. When you find that out, it's far too late.

relatively cool from the ice and cold of a long Midwest winter, studies show that cold fronts (known as backdoor fronts) are enhanced by the chilly lake water. These fronts push south and west of the lakes before settling over southeast Iowa and central Illinois. Once in place, they are reluctant to move because of the blocking effect of the cold, dense air flowing off Lake Michigan.

Eventually, warm moist air will glide up and over the front, generating rounds of thunderstorms that tend to move slowly and produce large amounts of rain. Lake effect fronts are ideal catalysts for extreme weather and are responsible for about one-third of our severe weather events during the summer months.

Without a doubt, Iowa's worst flash flood occurred in the tiny town of Rockdale, situated on a creek in a narrow valley just west of Dubuque. On July 4, 1876, after 5 inches of rain fell in a short period of time, a wall of water surged through the narrow valley on which Rockdale was situated, drowning 40 people.

Flash floods are dangerous in the relatively flat terrain of

Record Crest (feet)	Date	River Level
1	July 7, 1993	22.63
2	April 28, 1965	22.48
3	April 25, 2001	22.33
4	March 10, 1868	22.00
5	April 20, 1997	19.66
6	June 27, 1892	19.40
7	April 26, 1969	19.30
8	October 7, 1986	19.22
9	May 9, 1975	19.16
10	February 22, 1966	19.00

TEN HIGHEST MISSISSIPPI RIVER CRESTS AT LOCK AND DAM 15 IN DAVENPORT/ROCK ISLAND

Courtesy of U.S. Army Corps of Engineers.

Normal Mississippi River Stages for Each Month of the Year at Selected Lock and Dam Sites

Stage (feet)

CITY	LD	Jan	Feb	Mar	Apr	May	Jun	Jul	Aug	Sep	Oct	Nov	Dec	Annual
Dubuque	LD11	5.6	5.7	7.5	10.8	9.7	7.5	6.6	5.6	5.5	5.7	6.2	5.8	6.8
Bellevue	LD12	6.1	6.3	8.4	11.6	10.5	8.1	7.0	5.6	5.5	5.7	6.6	6.1	7.3
Clinton	LD13	5.6	5.5	7.6	11.2	10.2	7.8	6.7	5.5	5.5	5.6	6.3	5.7	6.9
Le Claire	LD14	5.0	5.0	5.9	7.9	7.3	6.0	5.5	4.9	4.8	4.9	5.3	5.0	5.6
Davenport	LD15	6.2	6.0	7.9	11.1	10.1	8.0	7.0	5.6	5.4	5.7	6.4	6.1	7.1
Muscatine	LD16	4.9	4.9	6.9	10.6	9.7	6.9	5.9	4.6	4.4	4.6	5.2	4.8	6.1
New Boston	LD17	5.9	5.9	8.3	11.8	10.6	8.2	7.1	5.2	4.7	5.1	6.0	5.7	7.0
Burlington	LD18	2.9	3.1	5.2	7.9	7.1	4.9	4.1	2.7	2.3	2.5	3.2	3.0	4.1

Courtesy of U.S. Army Corps of Engineers

the Midwest, but in mountainous terrain they can be life-threatening because the walls of the mountains or canyons do not allow the water to spread out. Instead, the water gets deeper and swifter and ultimately becomes a raging torrent. Most of our nation's worst flash floods have happened in mountainous areas where large amounts of rainfall created flash floods. In Johnstown, Pennsylvania, a flash flood killed 2,200 people when it roared out of the Appalachian Mountains and into the heart of the city in 1889. In 1972, a similar flood killed 238 people in Rapid City, South Dakota, when rainfall turned local creeks into boiling torrents of water that leveled buildings and trapped people in their cars.

In the Quad Cities, we live on the Mississippi. The mighty river is the second longest in the U.S. with a total length of 2,350 miles. It gets its name from the Algonquian Indians, who showed their respect by giving the river a name that meant the "father of waters." It plays a prominent role in the lives of those who choose to live, work, and play along its path. The river's moods are varied. In winter, it turns slower, and is often choked with ice from December through February. Come April, snowmelt and rains cause significant rises, and, in the spring, the river crests. Some years, when those variables come together just right, the river becomes dark, swift, and dangerous, climbing 10 to 15 feet above normal. Then come the dog days of summer and river levels drop and the Mississippi returns to its slow-moving, sleepy self.

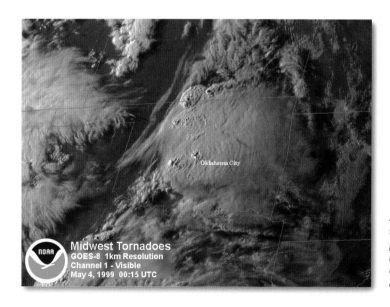

A satellite image of a mesoscale convective complex that produced deadly tornadoes over Oklahoma City, Oklahoma, in 1999.
COURTESY OF NOAA

The Great Flood of 1965

There is a place on the Mississippi River where the water inexplicably changes direction. Heading west instead of south, the river cuts through hills lush with trees and flows through a metropolitan area known as the Quad Cities. The community was created because of the river, and the river is its soul. The river is revered and celebrated for the wonder that it is as it moves at a leisurely pace through the lives of Quad Citians. On occasions, however, the river can be fickle, moody, and downright angry.

Until 1993, the Great Flood of 1965 was the measuring stick for all other floods on the 2,350-mile Mississippi River. For most of April, the river poured out of its bank before reaching a record crest in the Quad Cities on April 28, 1965. Nearly 40 years later, the flood of 1965 has a legendary place in the Quad Cities' weather history.

For a major flood to happen, a number of meteorological events line up in just the right way. William Joern, the head of the National Weather Service in the Quad Cities at the time of the 1965 flood, says major floods are dependent on luck—call it good luck or bad luck—"just like a Las Vegas gambler coming up with three bars on a slot machine."

The first inkling of trouble in the Quad Cities came on March 19, 1965 when the U.S. Weather Bureau warned areas along the Mississippi that the river might exceed flood stage as far south as Quincy, Illinois. The statement was based on the heavy snowpack and unusually deep frost levels over much of the Upper Midwest. People were concerned, but nothing yet pointed to a flood of historic proportions.

That changed in early April when widespread heavy rain fell. Because of the combination of snowmelt, rain, and

Raindrops Keep Fallin' on My Head

In drawings, raindrops are often shaped like teardrops. In actuality, raindrops bear scant resemblance to tears. Small raindrops are circular, while larger ones are round at the top and flat on the base, like a hamburger bun. As raindrops grow, the biggest ones take on a parachute shape with a narrow tube of water near their base. These raindrops eventually break into smaller drops.

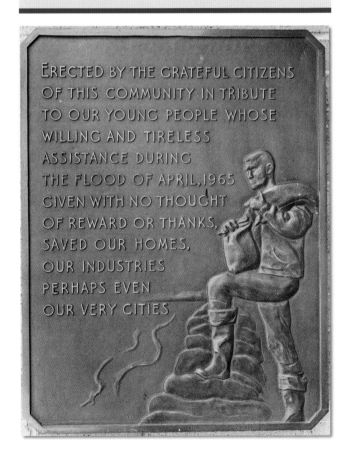

This bronze bas-relief marker on the Iowa-Illinois Centennial Bridge is a tribute to young people who helped sandbag during the April 1965 flood. It features Ken Becker, a 17-year-old Bettendorf flood fighter.

frozen ground, large amounts of water accumulated in the Mississippi River and in the surrounding tributaries. With nowhere to go but up, the river began to expand.

On April 8, 1965, the U.S. Weather Bureau issued a flood advisory for much of the Mississippi River watershed, including the Quad Cities. At Lock and Dam Number 15, people were shocked when the bureau projected the river's crest to be 19.5 feet. The river had never, in its recorded history, been measured at that level. The previous record crest of 19.4 feet, set in 1892, was likely to be broken in just two to three weeks. With the dire news, flood preparations shifted into high gear.

Back in 1965, few towns on the Mississippi had flood-walls. To keep muddy floodwater out of the towns, people built sandbag barriers or moved dirt into makeshift walls called dikes or levees. Time-consuming and labor-intensive, these structures became battlefronts manned day and night by thousands of volunteers. As the scale of the flood war escalated, so too did the need for warriors. To meet the demand in the Quad Cities, almost every college and high school student was excused from school to assist with sand-bagging. Working in rainstorms and chilly spring conditions, volunteers stood in long lines and passed 30- and 70-pound sandbags along until they reached the last volunteers, who placed each one in the levee.

During the 1965 flood, one of the biggest battles took place at the Moline, Illinois, waterworks. In an emergency meeting on April 11, 1965, Moline Mayor James Arndt stated that protecting the waterworks was a top priority because he knew that the projected crest would submerge the waterworks and contaminate city water. He planned to save the waterworks by having city workers construct a floodwall out of dirt, and then shore it up with 30,000 sandbags.

Just as the work on the floodwall got underway, the Weather Bureau had more bad news. Widespread heavy rain was expected, which would raise the level of the river. Based on these new forecasts, the Mississippi River was expected

to crest at 20.5 feet on April 16, 1965. The rains were heavier and more widespread than expected, so another revision was made on April 22, 1965 for a crest of 21.5 feet several days later.

A severe flood was now a certainty.

On April 24, the Mississippi River crested at Guttenberg, Iowa. The river surged through the bluffs of northeastern Iowa, heading south to Dubuque. As

Weather Folklore

If the muskrat move its nest away from the water's edge, you can expect a flood.

Volunteers fill sandbags to shore up dikes and levees on the Mississippi River.
PHOTO COURTESY OF *QUAD-CITY TIMES*

77

What to Do During a Flood Warning

- Follow the instructions for a flood watch.
- Listen to local authorities. Local authorities are the most informed about flood affecting your area.
- If you think you are at risk, evacuate immediately. Move quickly to higher ground. Save yourself, not your belongings.
- If you are advised to evacuate, do so immediately. Move to a safe area before rising water becomes too deep and your escape route is cut off.
- Follow recommended evacuation routes. Shortcuts or alternate routes may be blocked or damaged.
- Leave early enough to avoid becoming marooned on flooded roads. If you delay too long, escape routes may be blocked.
- Stay out of areas subject to flooding. These include dips, low spots, canyons, and washes that can become filled with water and flood.
- If you are outside, climb to high ground and stay there. Move away from dangerous streams and rivers.
- Never, ever, try to walk, swim, or drive through swift water. If you are walking and come to a stream where water is above your ankles, stop, turn around, and go another way. Even water 6 inches deep can sweep you off your feet if it is moving swiftly enough.

What to Do During a Flood Watch

- Floods and flash floods happen quickly, and without warning, so everyone in an affected area needs to be prepared to act quickly.
- If you live in a flood-prone area, be aware of any places where streams or creeks have flooded. Be ready to evacuate at a moment's notice.
- Listen regularly to a NOAA Weather Radio, which provides a continuous weather updates, or listen to a portable, battery-powered radio or television, for emergency information. Local radio and television broadcasters offer the most immediate, up-to-date information about an area. They will be able to tell residents which areas are most affected and what emergency precautions to follow.
- Follow the instructions and advice of local authorities.
- Fill bathtubs, sinks, and plastic bottles with clean water. As the floodwaters rise, water may become contaminated and water service may be interrupted.
- Bring outdoor belongings, such as patio furniture, indoors. Unsecured items may be damaged or swept away by floodwaters.
- Move your furniture and valuables from the ground level or basement of your home to a higher story. In the event your home is affected by a flood, the higher floors are less likely to be damaged.
- Locate the main power switch and gas. In some areas, local authorities may advise you to turn off utilities to prevent further damage to homes and the community.
- Get your disaster supplies ready. You may need to act quickly. Having your supplies ready will save time.
- Fill up your car's gas tank so you are prepared to travel if an evacuation notice is issued. If electricity is cut off, gas stations may not be able to operate pumps for several days.
- Be prepared to evacuate. Local officials may ask you to leave the area and head to an evacuation center if a severe flood is threatening your home.

Water swamping Fulton, Illinois. COURTESY OF QUAD-CITY TIMES

water from tributaries throughout the Upper Basin of the Mississippi poured into the river's main stem, the National Weather Service's Flood Forecast Center issued its final, even more alarming, advisory for the Quad Cities. The river was expected to crest at 22.5 feet—a foot higher than the previous prediction—four days later on April 28, 1965.

Meanwhile, buckets of rain soaked soggy sandbags, dikes, and levees as volunteers struggled to work against time. In East Moline, 850 families were forced to leave their homes. The East Moline Water Plant was forced to shut down, while city officials and volunteers struggled to keep Moline's waterworks from experiencing the same fate. The city of Moline began supplying water to East Moline residents as it was available. Mayors from both cities urged people to limit water use, which included forgoing bathing and washing clothes.

The river's crest passed Dubuque and surged rapidly southward. A river crest can be tracked by measuring the river's height at a determined spot. When measurements show the river rising, the crest is still approaching. When it begins to fall, the crest has passed. The highest level measured is considered the actual crest. By Monday, April 26, the water broke a dike at Lock and Dam 13, covering Illinois Highway 84 in six spots, encircling the town of Fulton, Illinois and turning it into island. Most of the eastern edge of Clinton was also submerged. The western part of Clinton was high enough to avoid flooding.

By April 27, record river crests had been established in the Iowa towns of Princeton and Le Claire along the river, as well as the Illinois towns of Port Byron, Rapids City, and Hampton. When the water topped Lock and Dam 14, a dangerous break was discovered in the 1,000-foot levee on the Illinois side. If the levee failed, much of Hampton would be under several feet of water. Mayors and other designated officials issued an urgent plea for help, and within two hours, 100 volunteers arrived on the scene, along with truckloads of gravel and sandbags. Eventually, 600 tons of

What to Do If You Are in a Car

- If you are in your car when you hear a flood is in your area, check your radio or television to find out what areas are flooded and avoid them.

- Do not cross streams of water on the road because the depth of the water is not always apparent. The roadbed may be washed out under the water, and you could be stranded or trapped.

- Most flood fatalities happen when people attempt to drive through floodwater. Rapidly rising water may stall the engine or engulf you and your vehicle and sweep you away. Look out for flooding at highway dips, bridges, and low areas. Two feet of water can carry away most automobiles.

- If you are driving and come upon rapidly rising waters, turn around and find another route. Move to higher ground away from storm drains, creeks, streams, and rivers. If your route is blocked by roadside barricades, don't risk your life by trying to drive through them. Turn around and find another route. Barricades are put up by local officials to protect people from unsafe roads.

- If your vehicle is surrounded by water or the engine stalls, abandon your vehicle and climb to higher ground if you can safely do so. Many people die during floods because they attempt to move stalled vehicles. When a vehicle stalls in the water, the water's momentum is then pushing against the car. As the water rises up, the car's weight is displaced and the car becomes buoyant. One foot of water moving at 10 miles per hour exerts a lateral force of about 500 pounds on the average automobile. Two feet of water moving at 10 miles per hour will float virtually any car.

- If you have to abandon your vehicle, use caution. Look for a chance to move away quickly and safely to higher ground. Many people have been swept away by streams upon leaving their vehicles, which are later found without much damage.

gravel and 10,000 sandbags were added to the levee at "Fisherman's Corner" north of Hampton, Illinois. In the end, despite the fact that the river trickled across its top, the levee held.

By April 28, the crest was coming to call on the Quad Cities. As river water poured into the heart of the town, hundreds of volunteers continued to sandbag and man the shaky dikes and levees. In one last stand, weary residents watched as the swollen river licked and chewed on their handmade floodwalls. The river reached rose and, with no more than a rush of current, it crested.

At Lock and Dam Number 15, the gauge read 22.48 feet. This newly established crest was an all-time record, easily breaking the previous mark of 19.40 feet set in 1892. The river remained at that height for a day, until the water levels began to recede on Friday, April 30, 1965.

As the floodwaters receded, the hopes of volunteers on the levees soared. In Moline, the weeklong battle to save the waterworks was declared a success. Throughout the city, the walls of sand and dirt had held and only a handful of families were forced to leave. For those on the frontline, it was time for a shower and some long awaited shut-eye.

East Moline was not as lucky. Hundreds of people were evacuated to shelters or to the homes of family and friends. The subdivisions of Cottage Grove and Watertown were hardest hit. On Campbell's Island, water nearly 4 feet deep flowed into many homes. When the flood was at its peak, the members of the East Moline Police force worked 12-hour days to protect the city and assist flood victims. When the flood was over, the police force, in an exceptionally generous gesture, donated their overtime to the community.

In Rock Island, Illinois, the 350,000 sandbags piled 6 feet high pretty much protected the city—only 30 homes were damaged by the flooding. When a dike failed at the Rock Island Boat Club, however, nearly 1,200 people were evacuated from Arsenal Courts housing project. As a result, the J.I. Case plant that manufactured farm equipment in Rock Island was forced to shut down.

In some way, shape, or form, everyone who lived in the Quad Cities was affected by the flood. In Davenport, Iowa, the flood-prone Garden Addition was a maze of sandbags and pumps and two hundred families were evacuated. Le Claire Park, Credit Island, River Drive and the

Staying Put: Silent Henry

There were those who fled reluctantly before the great Mississippi flood of 1965. There were those who cursed the river. There were those who forgave the mighty stream its trespasses. There were also some who stayed put, despite the danger.

One man became legendary, despite his

"Silent Henry" rides out the flood

"Silent Henry," his face obscured by the smoke from his woodstove, rides out the 1965 flood on his makeshift raft.
COURTESY OF *DAVENPORT TIMES-DEMOCRAT*

efforts to shun publicity. The man Clinton residents dubbed "Silent Henry," was a Noah-like figure of few words who rode out the Mississippi's mightiest flood alone on a tiny raft moored to a tree that rose from a flooded island. The raft was located directly over his water-covered shack. As the river rose, so did Silent Henry's raft.

The man, whose real name was Henry Steele, was 72 years old and had lived on Clinton's north end for nearly 30 years. He refused all offers of help and brandished a shotgun at those who approached his bobbing raft, on which a small fire in a barrel stove smoldered continuously. According to the legend, he subsisted on mud hens, fish, and river water, and it wasn't unusual to hear his shotgun blasts in the middle of the night, far out on the swollen river.

old Municipal Stadium were destroyed, and bridges were closed to traffic. Arsenal Island could not be reached by the bridges that connected it to Davenport or Moline, Illinois. Throughout the metropolitan area, traffic was disrupted by numerous street closures and detours.

In the 28 days that the river was above flood stage, over 11,000 Quad City residents were evacuated from their homes. In Rock Island County, the Red Cross provided overnight shelter for more than 14,000 people and served approximately 45,000 meals. Were it not for the efforts of hundreds of these volunteers, young and old, the numbers would have been significantly higher.

Nobody knows exactly how many sandbags went into the frenzied fight to keep the river water from inundating eastern Iowa and western Illinois, but according to the U.S. Army Corps of Engineers in the Rock Island District, 3,600,000 bags were issued from Dubuque, Iowa, to Hannibal, Missouri. Davenport and Clinton threw down 200,000 bags each, and Dubuque reported that 350,000 sandbags were used to bar the river.

If the Great Flood of 1965 had a legacy, it is the floodwalls. During the flood, it became apparent that the major cities that reside on the Mississippi's banks needed floodwalls. In the years that followed, a great deal of time and effort was spent on creating the walls that now line much of the riverfront in the Quad Cities. Although citizens in Davenport have remained steadfast in their desire to remain free of floodwalls because they block the picturesque views of the river, the towns of Bettendorf, East Moline, Moline and Rock Island have erected floodwalls.

Today, the big river still cuts a path through the core of the Quad Cities. New floods have come and gone. During the flood of 1993, river levels even exceeded those of the flood of 1965. But because of the lessons learned during that long, hard April, other floods have not been as disruptive or damaging to the people and property of the Quad Cities.

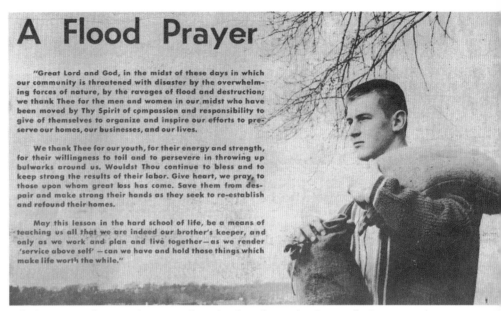

A Flood Prayer

"Great Lord and God, in the midst of these days in which our community is threatened with disaster by the overwhelming forces of nature, by the ravages of flood and destruction; we thank Thee for the men and women in our midst who have been moved by Thy Spirit of compassion and responsibility to give of themselves to organize and inspire our efforts to preserve our homes, our businesses, and our lives.

We thank Thee for our youth, for their energy and strength, for their willingness to toil and to persevere in throwing up bulwarks around us. Wouldst Thou continue to bless and to keep strong the results of their labor. Give heart, we pray, to those upon whom great loss has come. Save them from despair and make strong their hands as they seek to re-establish and refound their homes.

May this lesson in the hard school of life, be a means of teaching us all that we are indeed our brother's keeper, and only as we work and plan and live together—as we render 'service above self' — can we have and hold those things which make life worth the while."

A flood prayer, given by Reverend Emerson Miller, with a photo of Ken Becker, the young flood fighter who became a symbol of the 1965 flood.. COURTESY OF *DAVENPORT TIMES-DEMOCRAT*

The great flood of 1965 inundated homes and neighborhoods throughout the Quad Cities.
PHOTO COURTESY OF *QUAD-CITY TIMES*

— Tranquil Stream to Raging Torrent: The 1990 Duck Creek Flood —

WASHOUT!

The flash flood of June 16, 1990, caused terror and tragedy for thousands of Quad-Citians. It also brought out the triumph of the Quad-City spirit.

This dramatic rescue during the Duck Creek Flood occurred when a Davenport man, his blind wife, and their two children were pulled from their car minutes before it sank.
COURTESY OF
QUAD-CITY TIMES

just as challenging as sounding the alarm. From my home to the station, numerous roads were closed and under water; finding an open route was a monumental challenge.

The story of the flood began in late spring 1990. A soggy May had dumped 7.68 inches of rain on the Duck Creek basin. June thunderstorms brought the six-week rain total to more than 10 inches. Two weeks into June, the ground was saturated, and farmers were grumbling. Quad Citians were beginning to long for a stretch of warm, dry weather.

When Friday, June 15, dawned sunny and dry, spirits soared. As the day unfolded, temperatures climbed to a high of 84 degrees. The humidity had also increased, but by June standards, it was still a fine late-spring day

Changes, however, were afoot. To the south, in Missouri, temperature readings cracked the 90s. To the north, cool high pressure had set up shop over the Great Lakes, which limited temperatures to the 60s and 70s in that region. A distinct thermal boundary, enhanced by the cool water of Lake Michigan, had established itself just south of the Quad Cities. On the weather charts,

Many storms have come and gone in the Quad Cities, but few have been as intense, furious, or destructive as the 1990 Duck Creek Flood. In just a few hours, a narrow, tranquil stream became a raging torrent nearly a mile wide.

For the people of the Quad Cities, the storm was a frantic time when lives, homes, and businesses were damaged. For me as a weatherman, the storm was a defining moment—a moment when the lives of Quad Citians depended on the efficiency, clarity, and accuracy of my work. Getting out storm warnings meant getting to work, and that task proved to be

The Speed of Raindrops

Raindrops come in different sizes, which affects the speed at which they fall. Without a wind present, a raindrop the size of a housefly hurtles to the ground at 20 miles per hour. A drizzle-sized drop falls at 5 miles per hour. The average free-falling skydiver, by comparison, falls at 125 miles per hour.

the boundary resembled a wall that separated the coolness of spring from the heat of summer.

At the intersection of that front, thunderstorms were likely where the atmosphere was deceptively unstable. I was confident thunderstorms would form, and that night in my forecast, I indicated some were capable of producing heavy rains. Knowing the ground was soggy, I mentioned the threat of run-off problems. Flash flooding crossed my mind, but only minimally. After a Friday night, post-work pizza, I called it a day.

Less than an hour later, I awoke to lightning. The storms I had expected had arrived. I was pleased. Always captivated by lightning, I moved to a chair, where I watched it flicker over the silhouettes of neighborhood houses.

I awakened two hours later when a loud clap of thunder rattled the windows. Rain cascaded down the side of the house, and I could see the wind bend the maple tree that I had recently planted. Sheets of rain fell. The little creek behind my house was becoming a stream. Unabated, the rain continued. I watched in amazement as the water reached the basement next door.

Then the storm diminished. The water receded. One more time, I went back to bed. This time, however, I felt uneasy. New lightning had begun to flash in the west.

The sleep I had hoped for never came. New thunderstorms formed as the stationary front guided the individual storms like trains on a track. One after the other, they rolled over the same general area, a weather phenomenon known as "training." By all accounts, this was a textbook case.

I realized, with a start, that flash floods were no longer a possibility—they were a given. For the first time, I sensed the magnitude of what was taking place.

At this point, the next wave of storms blew through, as furious as the last, but the sky was getting light. The rain had been falling nonstop for nearly four hours. I threw on a T-shirt and gym shorts, frantically washed my face and brushed my teeth, and headed for the television station.

This fireman wades through flood waters to assist a stranded motorist.
PHOTO COURTESY OF *QUAD-CITY TIMES*

This map illustrates the heavy rainfall of June 16, 1990 that precipitated the flood.
COURTESY OF *QUAD-CITY TIMES*

Rescues of stranded motorists, family pets, even a family sofa occurred when cars, homes, and entire neighborhoods were engulfed by water.
COURTESY OF
QUAD-CITY TIMES

lapped at car doors. People stood in crowds, pointing as chairs, toys, pots, and pans went racing by. Others, some dressed in pajamas, looked on in horror as muddy water poured into their homes.

This, I realized with a shock, was a real-life disaster. I had seen pictures of floods, but I had never expected one in my backyard. Heart pounding, I threw the car into reverse, dodging the other stunned drivers. After driving around, I finally found a route that got me over Duck Creek and down to the station. It took me nearly an hour to navigate the 7 miles from Bettendorf to KWQC—a ride that usually takes 15 minutes.

As I raced in the station door, the sky looked brighter and the rain was subsiding. The water, nevertheless, was still rising. It was around 8 A.M., and as

I was so rushed I didn't even think to take a suit and tie. Time was of the essence.

As I pulled out of the driveway, I didn't know that just 11 miles away, at the headwaters of Duck Creek, 10 inches of rain had fallen. Because the water could not drain through the city's concrete, and the ground was saturated, the water took the path of least resistance: Duck Creek. Usually a tranquil backdrop for bikers and joggers, Duck Creek was swollen and boiling and still growing. Muddy and full of debris, the creek carved a trail of destruction as it roared through the middle of the two sleepy towns of Davenport and Bettendorf.

After a few minutes of driving on roads narrowed by small-scale lakes, I came to Middle Road. I could not believe my eyes. A quarter-mile-wide river of water poured across the road. Traffic in both directions was stopped. Water

it turned out, I would not leave the building again until after midnight.

My first concern was getting on the air. Because it was Saturday, there were no morning broadcasts, which meant gathering up people with the technical skills to get me on the air. I called in a team of two anchors, several reporters, a director, and a camera person to cover the fast-breaking developments.

About half an hour later, with a skeleton crew, I finally hit the air.

We put together a special live newscast to keep viewers up-to-date on the rapidly changing conditions. While I talked about the weather, the anchors covered gas leaks and emergency procedures. They reported that roads were blocked all around town, people were stranded, the power was out, and emergency crews were responding as best they could. Helicopters were being called in to perform rescue operations. I remember how odd it felt to be on the air in a dirty T-shirt and tennis shoes.

By that evening, the rain had subsided and Duck Creek was once again flowing between, and not over, its banks. Four people had lost their lives in the flood. Nearly 8,000 homes and businesses sustained damages estimated at more than $25 million. The National Guard was called in to assist. Terry Branstad, the governor of Iowa at the time, came to town and declared what was obvious to locals: Duck Creek was a disaster area. The 5 to 10 inches of rain that had fallen in six hours' time had turned the town into a quagmire

When I left the station that night—nearly 14 hours after I had arrived—I drove as close to Duck Creek as the barricades would allow. When I could go no farther, I walked into the normally quiet neighborhoods that were now chaotic. The steady chugging of pumps and the roar of generators filled my ears. Yards were filled with furniture and appliances, rugs and carpets were stretched across chairs and couches, anything that would support them. Windows and doors of homes were wide open, showing the shells of water-stained walls. Out in the streets, debris and sludge made walking a challenge. I saw the mud-streaked faces of the residents, shocked as they surveyed their lost homes and belongings. These things, however, could be replaced. The lives of the four Quad Citians swept away by the flood could not.

Minutes later, a National Guardsman shone a flashlight in my face asked me if I had a reason to be there. I looked down at my mud-caked shoes, shook my head, and walked away.

Experts say another 10-inch rain may not occur for a hundred years. To this day, however, when we get one of those storms where the rain comes down in sheets and lightning skewers the sky, I always think of the Duck Creek Flood of 1990.

Saving lives, homes, automobiles—and even the family sofa—were priorities during the Duck Creek Flood. COURTESY OF *QUAD-CITY TIMES.*

Law enforcement agents worked to prevent looting and keep sightseeing to a minimum. COURTESY OF *QUAD-CITY TIMES*

The Ring of Fire and Water: The Great Flood of 1993

I've been though more than my share of floods, but I was not prepared for the onslaught of water that was the Great Flood of 1993. In a three-month period, the summer floods grew in scope from a minor inconvenience to an historic disaster. Working the event had a similar pattern. Our coverage started small and, with time, grew from casual outlooks on the river's flood potential, to critical information on river stages, rainfall predictions, and eventually—when the river swelled to staggering proportions—flood warnings.

It was a stressful time highlighted by presidential visits and national television coverage. The eyes of the world were on Iowa. It was a time for communities to work together to fight a battle that was started by the forces of nature.

As one of the voices of that epic flood, I found great sat-

This welcome sign greeted visitors to Davenport's downtown during the height of the 1993 flood. COURTESY OF QUAD-CITY TIMES

isfaction in being a positive force in the struggle. Despite often being the bearer of bad news, I was still able to provide straight talk and timely information on which crucial decisions were based. Even before record crests, I knew full well I was a part of something that would be talked about for generations to come. It strengthened my resolve to raise my performance to the highest level possible.

After the Great Flood of 1965, when the Mississippi River rose to record levels, the river had its ups and downs, but for the most part it was a haven for boaters and fisherman, and a travel route for the big-bellied barges regularly churning back and forth from the Gulf of Mexico to the shores of the Twin Cities. Residents along the river were lured into a false sense of security.

Then came the Great Flood of 1993, when the Big Muddy once again flexed its muscle and residents were reminded about the power of its mighty waters.

REPRINTED BY PERMISSION OF JACK OHMAN/THE OREGONIAN

The story of the flood begins in the fall of 1992, when frequent rainstorms soaked the Midwest. In Iowa, Minnesota, and Wisconsin, November 1992 was the wettest on record since 1895. With winter, the rain turned to snow, especially in the North. By the end of February, snow 9 to 18 inches deep and with a water content of 2 to 4 inches covered the southern halves of Minnesota and Wisconsin. Farther north, near the headwaters of the Mississippi in northern Minnesota, the snow was measured in feet.

As quietly as snow falling, the conditions necessary for spring floods were falling into place. The ground was saturated, and the snowpack was deep. With a quick thaw and heavy spring rains, the threat would become a reality.

As luck would have it, heavy rain and cool temperatures gripped the region that spring. Many areas, including the Quad Cities, received more than 9 inches of the wet stuff in April and May. The heavy precipitation, combined with water from the melting snowpack up north, saturated the ground.

Hydrologists warned people in the floodplain to fill sandbags and move to higher ground. The river was in a mood, and there was nothing anyone could do but wait for the crest of water that was sure to come.

On March 23, 1993, high water swept down the river from the bluff country of Minnesota to the Quad Cities. The river rose nearly 2 feet in 24 hours, from 6.8 feet to 8.54 feet. During the ensuing days, the river climbed, reaching the 15-foot flood stage early on April 7. For the next three weeks, the river remained at flood stage as it boiled through the heart of the Quad Cities. On April 25, 1993, the crest of 18.6 feet at Lock and Dam Number 15 was the fourteenth highest crest in Quad-City history. The river began to recede.

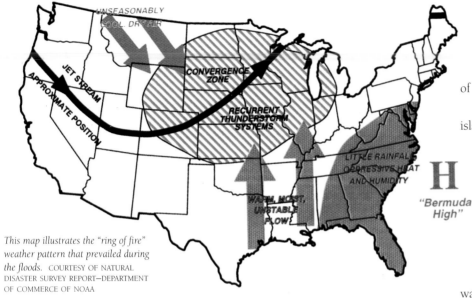

This map illustrates the "ring of fire" weather pattern that prevailed during the floods. COURTESY OF NATURAL DISASTER SURVEY REPORT–DEPARTMENT OF COMMERCE OF NOAA

88

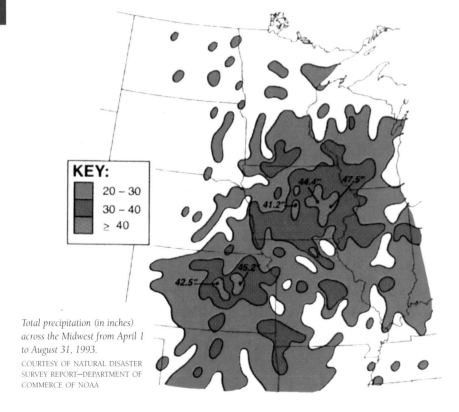

KEY:
- 20 – 30
- 30 – 40
- ≥ 40

Total precipitation (in inches) across the Midwest from April 1 to August 31, 1993.
COURTESY OF NATURAL DISASTER SURVEY REPORT–DEPARTMENT OF COMMERCE OF NOAA

People along the river were relieved. The crest caused by the annual spring snowmelt was out of the way. The worst was behind them. On the succeeding days, the river inched down, and the cool, wet spring showed signs of letting up. Life was getting back to normal.

Then came the ring of fire. Two years before on a distant island in the Phillipines, Mount Pinatubo erupted, nine times in all, climaxing on June 15, 1991. The eruption discharged millions of tons of sulphur dioxide in the earth's atmosphere that increased worldwide temperatures for years to come. While it's tough to measure and difficult to comprehend, this volcanic eruption in the Phillipines may have been a catalyst for the 1993 floods in the Midwest.

After the eruption, a massive high pressure system was born on the island of Bermuda. The creation of a high pressure system in Bermuda was typical for this time and place. The strength and endurance of this system, however, were anything but normal. Like a boulder in a creek, the high pressure diverted the jet stream so that moisture from the Gulf of Mexico flowed, unabated, into the heart of the country. In the eastern United States, under the high pressure system, it was hot and dry. On the jet stream's periphery, right over Iowa, the heat encountered cooler air from Canada. This clash zone, known to meteorologists as the

A Word In Time

If you live or play in an area that is subject to flooding, there is information available to keep you up to date on water levels. Advance Hydrologic Prediction Services was created after the 1993 floods to minimize the loss of life and property from flooding. Using computer models, automated gauges, satellites, and Doppler radar, meteorologists create forecasts for all of the major rivers. Get the latest information at www.crh.noaa.gov/dvn/ahps/.

"ring of fire," would become the breeding ground for intense rounds of thunderstorms and flooding rains.

As these disturbances slid along the front, they forced warmer air from the south to ride up and over the cooler air to the north. In this unstable atmosphere, warm air rose up and condensed into towering thunderstorms. As the thunderstorms moved along the jet stream, they encountered a stationary boundary that stretched across the Upper Midwest and they began to cycle over and over, soaking the same soils in the process known as training. For hours at a time, individual storms marched over the same general region producing exceptional rainfall totals.

Late in May, when the worst of the wet weather is typically over for the Midwest, the ring of fire was securely anchored. Throughout the Midwest, rain fell every single day from June through early August.

In the Quad Cities, the first day of June was miserable. The morning low was a record-breaking 39 degrees, with intermittent clouds and showers; the high never climbed out of the 60s. June 4 was even worse—an inch of rain poured down and the high was just 56 degrees, about 25 degrees colder than the average. Crops languished in the cool, muddy fields. In town, however, the residents' rush to cut the thick, green grass between rainstorms was a never-ending task.

After a brief reprieve, the rains returned. On June 7 and 8, thunderstorms pounded the Quad Cities and swamped the soggy soil with 2 to 3 inches of water. The Mississippi, which had shrunk to the 11-foot mark, was rising again, reaching the 15-foot flood stage on June 11.

Summer was just a week away and the grand old river was at unprecedented levels. For a few days, the river held steady. Residents hoped the hotter season ahead would force the storm to track north, along with the monsoon-like rains that had soaked the region for months on end for end. For those along the 2,350-mile length of the river, a critical moment in history lay just ahead.

In many areas, sandbags were not enough to prevent the Mississippi River from overflowing its banks.
PHOTO COURTESY OF *QUAD-CITY TIMES*

On June 16, 1993, thunderstorms lit up the night skies of the upper Midwest, fueled by heat and moisture near a stationary front. For three days, the storms raged, dropping an average of 2 to 7 inches of rain on southern Minnesota, northern Iowa, and southwest Wisconsin. In Baraboo, Wisconsin, 12 inches of rain fell in four hours. This series of rainstorms, cascading onto the saturated ground and into the swollen tributaries of the Mississippi, became the catalyst for the Great Flood of 1993.

Mid-June was an unusual time of year for a flood. Prior to 1993, most major floods were caused by snowmelt were in the spring. While run-off contributed to higher-than-normal spring water levels in 1993, the primary cause of the flood was the numerous, intense thunderstorm events over widespread areas. According to experts, it was a meteorological

President Bill Clinton visited the flooded Quad-City area on July 4, 1993 and offered $1.2 billion in federal relief aid.
COURTESY OF
QUAD-CITY TIMES

After the new storms, hydrologists revised their crest figures to 19.6 feet on July 2. The Corps of Engineers decided to close the stair-step system of locks and dams on the river. Lock and Dam Number 17, at New Boston, Iowa, was the first to fasten its gates, and barge traffic came to a halt. River front events were rescheduled, relocated or cancelled. The President Riverboat Casino pulled anchor and moved to the higher ground of Oneida Street Landing. Recreationists were told to stay away from the swelling waters.

On June 27, the river climbed to 18 feet. River Drive in downtown Davenport was closed as was River Drive in Moline. Sandbagging shifted into high gear on both sides of the river as volunteers made a last-ditch effort to keep water from businesses, homes, and streets. Meanwhile, new storms erupted, dumping an unwelcome 1.34 inches of rain on the Quad Cities. June 1993, with its total of 11.54 inches of rain, became the wettest month in recorded Quad-City history.

Monday, June 28, the river was on its way to 19 feet. The Rock Island Lock and Dam Number 15 was closed. Barge traffic was now at a complete standstill on a major stretch of the Mississippi. As the situation worsened, the Quad Cities Chapter of the American Red Cross formulated plans to handle the emergency.

On Tuesday, more storms increased the June rain total to an incredible 13 inches in the Quad Cities. The National Weather Service predicted the river would rise to 20.7 feet on July 2. Two hours later, they upped this to 21 feet. With the somber news, Scott County put out a call to the National Guard to help fill sandbags and to help keep looters away from freshly submerged areas.

Finally, June came and went, cloudy and wet. Two more inches of rain fell over the Upper Mississippi basin. When the new numbers were crunched, the result was astounding. A new crest of 22.5 feet was projected at Lock and Dam Number 15, making it the worst flood in history. Iowa Governor Terry Branstad toured the region and, along

phenomenon so unique that it may not occur again for another 500 years.

On June 23, hydrologists with the National Weather Service predicted that the river would rise 2.2 feet higher to an 18-foot crest on the Fourth of July.

The next day, more storms pelted the Quad Cities with 3.6 inches of rain. New London, in southeast Iowa, measured an astounding 9 inches total. Farm fields that had been swamped since early April were now flooded. Some farmers, after already replanting twice, threw in the towel.

with Illinois Governor Jim Edgar, declared the Quad Cities an official disaster area.

On July 2, the river rose to 21.70 feet. Davenport shut down its sewage system, and raw sewage flowed into the heart of the Mississippi. Anyone in direct contact with the contaminated water was urged to get a tetanus shot. The government bridge from Davenport to Arsenal Island was closed. The President Riverboat Casino was once again moved, this time to Bettendorf where it remained closed until the river receded. Lost business was costing the barge industry as a whole one million dollars a day. Curious onlookers crowded the Centennial Bridge that linked the states of Iowa and Illinois to the heart of the Quad Cities, trying to get a look at the swollen river and forcing local police to close the bridge to all pedestrians.

On July 3, when the river reached 22 feet—just a half foot short of the 1965 record—some hoped the river was

about to crest. Even so, hydrologists warned that flooding would continue for days to come.

On Independence Day, President Bill Clinton came to the Quad Cities to witness the unfolding disaster that was the Mississippi River. Standing on the Centennial Bridge

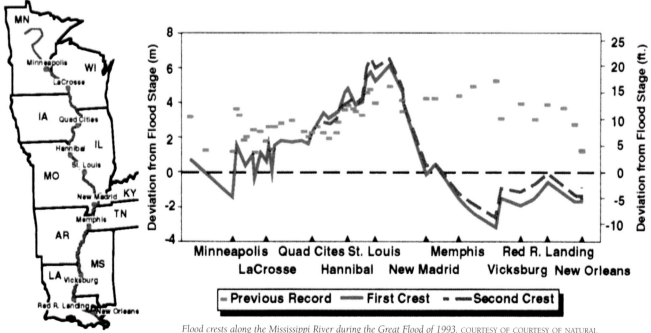

Flood crests along the Mississippi River during the Great Flood of 1993. COURTESY OF COURTESY OF NATURAL DISASTER SURVEY REPORT—DEPARTMENT OF COMMERCE OF NOAA

Kevin Hood wades through floodwaters that have overrun the sandbag levees.
PHOTO COURTESY OF *QUAD-CITY TIMES*

level of the river dropped, the spirits of the people soared. Then, more bad news: downpours had come to northern Iowa and Wisconsin to the north, which meant that a near-record crest was predicted for July 7. Tributaries such as the Iowa River and Cedar River in Iowa and the Rock River in Illinois were at record levels. Moline's popular Riverfest was cancelled.

The Mississippi continued to rise. All eyes were squarely on meteorologists like me who were frantically working the latest weather charts and satellites. We all came to the same conclusion: more rain was likely in the next two days.

Downstream, the Pope Creek levee near Keithsburg, Illinois, could no longer take the strain of holding back all that water. Boils that had signaled danger now grew into holes. Within minutes, the holes grew into a major breach and a wall of water blasted through Keithsburg and sliced the evacuated town in half. The nonstop sandbagging was all for naught. The city's water supply was quickly contaminated, and the muddy water, as deep as 13 feet in spots, engulfed 150 homes.

On Thursday July 8, the heavy rain predicted by forecasters arrived with a vengeance. A downpour rolled across the area and rapidly raised the river another 2 inches. At Lock and Dam Number 15, the water level reached 22.52 feet and finally, a record crest was established in the Quad Cities.

with the submerged streets of Davenport below him, he stated simply, "This is about as bad as it gets."

In the small town of Eldridge, Iowa, Clinton met with a large group of area farmers, who told him about crops that had failed to grow in the muck and mire of rain-soaked fields. Clinton promised $1.2 billion in federal aid to Midwest farmers. As his plane disappeared into the night sky, the river had climbed to a level of 22.03 feet—just inches short of breaking the record of 22.48 feet set April 28, 1965.

Over the next two days, as the

As rain inceased the level of the river, pressure mounted on the levees, dikes, and soggy sandbags. In the Garden Addition of Davenport, residents kept a constant vigil while National Guardsmen attempted to reinforce shaky floodwalls. In John O'Donnell Stadium, ducks patrolled the empty stands. Flooded roads and

METRO EDITION
Quad-City
Times
82
68
Drenching rains
Details, 4A
● FRIDAY, July 9, 1993

FLOOD WATCH: The Mississippi River at 22.6
50 CENTS

RIVER RISES INTO HISTORY

detours made navigating riverside thoroughfares demanding, if not impossible. The drone of pumps and the buzz of mosquitoes filled the air.

Finally, at 4 A.M. on Friday morning, July 9, the Mississippi made one last surge and climbed to 22.63 feet—an all-time record that still stands at Lock and Dam Number 15.

The river slowly receded—although it was above flood stage for three more weeks—leaving a layer of scum, debris, and garbage and a terrible stench in its wake.

Today, the Great Flood of 1993 is legendary. In terms of rainfall, river levels, flooding, crop damage, property damage, duration, the number of people displaced and the 38 people killed, the flood surpassed all others in modern United States history.

In the Midwest, major rivers combined to break 95 previous record flood crests, many by 6 feet or more. In the Quad Cities, the Mississippi was above flood stage on 83 of the 114 days between April 7 to July 30, 1993. On one 24-day stretch, the river in the Quad Cities was higher than any other crest in modern-day history, except that of 1965.

Hydrologists from the weather service estimate that in the Quad Cities, the flood is a 100-year event. Downstream in Burlington, a similar crest is not expected for another 200 years.

"Let me ask each of you to take heart and have faith," President Clinton said to the people of the Quad Cities in a radio address at the height of the flood. "As hard as these times are, you know that the waters will soon recede and the work of recovery will begin. The people who grow our food in the communities that surround and support you are central to the American way of life. Just as we depend on you for the harvest, you can depend upon us for support at this critical moment in your lives. For that is the American way."

The people of the Quad Cities survived and rebuilt. The Mississippi River, just as it has since the days of Mark Twain, runs through their hearts and unites them. To them, the Great Flood of 1993 was not so much a tragedy as it was a badge of honor for their courage, determination, and survival.

Volunteers, young and old, pumped water from flooded Davenport businesses.
COURTESY OF *QUAD-CITY TIMES*

93

CHAPTER
5

Heat Wave: The Sizzling July of 1936

Leo V. Brandmeyer of Rock Island, Illinois, took his last breath at 10:05 A.M. on July 6, 1936. The cause of his death was officially registered as heat—extreme heat. With his quiet passing that sizzling day, no one could have predicted what was to come. Yet, Mr. Brandmeyer's demise heralded the beginning of one of the deadliest natural disasters in Quad-City history.

To this day, the intensity and duration of that heat wave is unmatched. When it finally broke after two sweltering weeks, 72 people in the Quad Cities had perished, and the list of the dead filled a solemn column that ran the length of the *Davenport Democrat's* front page.

Today, those deadly two weeks and the hardships they inflicted are largely forgotten. Heat, after all, is an invisible foe. It does not strike with the speed and destruction of a tornado or flash flood. It is a relentless, silent killer that preys on the sick, the weak, or the elderly. It claims its victims one at a time.

Distant as it is, the July 1936 Heat Wave is the greatest this country has ever known and the one all others are measured against. It is remarkable not just for the number of people killed or the intensity of the heat it generated, but for the shear amount of real estate it cooked. All-time high temperature records were set in 15 states. Steele, North Dakota, for example, experienced an unprecedented high of 121 degrees—a temperature that was 181 degrees warmer than the reading of 60 degrees below zero it experienced earlier that winter. (This temperature variance in 1936 is the greatest in the history of the U.S. Weather Service.) The Great

95

DUST STORM APPROACHING SPEARMAN, TEXAS.
APRIL 14, 1935

RIGHT: COURTESY OF NOAA/HISTORIC NWS COLLECTION
FACING PAGE: *A searing sunset.* PHOTO BY MIKE HOLLINGSHEAD

Lakes wilted under the first 100-degree temperatures ever recorded in that part of the country.

Ironically, just a few months earlier Quad Citians would have paid a pretty penny for some of the warmth. The winter of 1936 had been harsh, even by Midwest standards. Snowfall totaled 45.9 inches, and February's average temperature of 11.6 degrees made it the second coldest on record. The average winter temperature of 16.4 degrees made 1936 the fifth coldest in Quad Cities history. On January 22, one of the worst days, the head of the Davenport weather bureau, T. G. Shipman, measured the temperature at a wind-whipped 22 below. In the *Davenport Democrat* he said that "it was the coldest day in 20 years."

As the calendar turned from February to March, the weather changed too. Even though it had been a time of drought and depression, the green grass and flowers buoyed the optimism of Midwesterners. As President Franklin Delano Roosevelt prepared to run for re-election, economic indicators pointed to signs that the Depression had run its course.

Silent and unobtrusive and nurtured by years of drought, heat was building over the Southwest. Like the air in a convection oven, it circulated, growing hotter with time. Finally, after feeding on itself for weeks, the heat moved north. Fueled by the jet stream, the worst heat wave in recorded history made its move on the Midwest.

Dog Days

The "dog days" of summer are those hot, steamy days of July and August. The term is tied to the constellation Canis Major (the big dog) and its brightest star, Sirius. Ancient Romans believe that Sirius, the dog star, was so bright it gave off heat. In midsummer, when Sirius would rise with the sun, the weather was hot and sultry and the sweltering period became known as the dog days.

It reached the Quad Cities on Sunday, the day after the Fourth of July. At the Federal Building, temperatures climbed from a low of 74 to 105 degrees by late afternoon. This was not only a record high temperature, it was just one degree lower than the hottest reading of 106 degrees set in June 1934.

Meteorologist T. G. Shipman said cooler breezes might return by Tuesday evening but that it was "a hope so remote" he did not include it in his weather forecast.

To escape the heat, people stayed in the shade, fanned themselves, and moved slowly. For $292, you could beat discomfort by purchasing a Kelvinator air conditioner available at the Mueller Lumber Company.

Air-conditioning was in its infancy, however, and few had the money for such a luxury. A pinch of salt in a cold glass of water was a recommended remedy. Bathing was also encouraged! An advertisement for Lifebuoy Soap entitled "Heat Wave On the Way" warned the reader not to "let the heat wave reduce you to an irritable sweltering pulp." Instead, the advertisement promised "you can be clean, cool, refreshed [and] free of the fear of B.O. (body odor), if you bathe regularly with Lifebuoy."

On July 6 and 7, Davenport experienced record highs of 105 degrees. On Wednesday, July 8, the Davenport temperature was 104 degrees. On July 9, it was 102

Sweat: The Body's Coolant

As early as 400 b.c., Hippocrates recognized that there was more to sweat than just water and salt. The stickiness we feel when the weather is hot is our body's way of trying to cool its core temperature. Each of us has approximately 3 million sweat glands that act as pumps, drawing water from nearby capillaries and routing it to the skin. This creates a layer of sweat, which evaporates and cools the body, thereby maintaining the body's normal temperature. And get this: scientists say sweat from heat is colorless as well as odorless!

Sweaty? Blame the Corn

Did you know that the crops planted in area fields can contribute significantly to the humidity in summer? According to agronomists, corn and soybeans transpire moisture in order to cool themselves. In the evening when temperatures cool, this moisture can be seen as a light fog hovering over farm fields. The moisture released in this process peaks in July and August, when the crops mature, and contributes to the hot, steamy weather known as the dog days.

Dust buried farms and equipment and killed livestock and people during the Dust Bowl years. COURTESY OF NOAA/HISTORIC NWS COLLECTION

97

98

degrees. Two more sweltering days of heat were promised, but there was reason for optimism. Shipman announced that "conditions causing the high temperatures are slowly losing their intensity."

Five days of excessive temperatures were beginning to take their toll. Peter Stein and Charles Orthman, both age 75, of Davenport, died at the Scott County Infirmary of "prostration." They passed away during the night when the temperature failed to drop below 82 degrees. Yards had turned from green to yellow and the grass became crunchy and sent up clouds of dust when walked on. Crops withered in the fields where the soil was split into gaping cracks from the relentless stare of the sun. Birds sang less. The pace of life slowed as the heat continued.

By Friday, July 10, the mercury tipped the scales at 105. The national death count reached 370 people. Meteorologist Shipman made a dramatic backpedal and declared that "all heat records will likely be broken within the next day or two. The conditions responsible for the excessive heat, and

which seemed to be losing intensity Thursday, have stiffened up and the worst is yet to come."

His words were prophetic. July 11 topped out at 107 degrees—a record for the day and the hottest temperature ever recorded in Davenport. This was also true for the next day, July 12, when the temperature at the Federal Building peaked at 108 degrees.

As the suffering increased, the *Davenport Democrat's* headlines proclaimed: "Relief near. Cool wave on way to baked areas." Shipman called for cooler conditions and thundershowers on Monday. After eight consecutive days of temperatures over 100 degrees, the worst appeared to be over.

Monday, July 13, dawned dry and hot. Relief was nowhere in sight. "Hope fades as rain clouds melt away," the *Davenport Democrat* stated. "Cool air unable to penetrate torrid zone of Midwest." By 2:00 P.M., the thermometer stood at 107 degrees. Another record had been shattered.

The nation had now lost 1,248 people to the heat, and the list was growing. In the Quad Cities, 18 had perished.

This included Minnie Huggins of W. Second Street in Davenport and Fred Lehns of Blue Grass.

The oven-like atmosphere was affecting everyone. Hot weather was, in the words of Des Moines District Judge Frank Shankland, "an enemy of happy marriage." He declared that "people are short-tempered, they flare up, file their petitions and get their divorce in a hurry."

People flocked to places such as Davenport's Le Claire Park, on the banks of the Mississippi River, where they slept outside in hopes of finding a cool breeze. Others looked for air-conditioned buildings to take the edge off. Walgreen's drug store was such a place, but business was down, as few people wanted to brave the heat just to get there. The Adler Theater in downtown Davenport had recently installed air-conditioning and allowed people to snooze on the floor for some sweat-free shut-eye.

As in all tragedies, one man's misfortune is another man's good luck. Around the Midwest, local merchants noted an upturn in the sales of cowbells. With pastures burning up, farmers let their cattle feed in forests so the poor animals could find some grass and some shade. At milking time, farmers had a difficult time rounding up their animals, unless the bovines were wearing the bells.

At midnight July 14, the thermometer measured 90 degrees. By daybreak it had cooled no farther than 84 degrees. Then the sun rose. By 8 A.M. the gauge read 94 degrees and at 10:00 A.M. it had climbed to 102 degrees. By 2:15 P.M., the temperature peaked at 111.3 degrees—133 degrees warmer that January 22, 1936, when thermometers registered the cold at 22 degrees below zero. That 111.3-degree reading remains the hottest temperature ever measured in the Quad Cities.

In Preston, Iowa, the large thermometer on the Maybohm garage burst into flame when the mercury expanded past the top. Without the mercury in the bottom of the tube, the exposed glass acted as a magnifying glass and set the garage on fire. Victims of heatstroke did not find much relief in hospitals, which were not air-conditioned. Doctors and nurses fought the heat around the clock in crowded, stifling emergency rooms.

By Wednesday evening, July 15, the temperature had risen to 106, and the death toll in the Quad Cities stood at 59, while the Iowa toll had climbed to 236.

Then, as quickly as it came, the heat wave left. On the morning of July 16, a cooling breeze descended on the area. Temperatures failed to reach the century mark for the first time in 11 days. Even though the worst was over, the list of victims continued to grow. "While the wave of torrid weather somewhat loosened its grip in the Tri-City area Saturday," the *Davenport Democrat* stated, "death did not do as much." By Sunday, July 19, the death toll in the Quad Cities was 72.

As the intensity of the heat faded, Quad Citians had to take stock of what had transpired. A whole community had been brought to its knees. Six dozen people had lost their lives. And the suffering was not confined to just humans, as large numbers of animals perished despite farmers' heroic efforts to provide water and shelter. Crops wilted, and yields were down, another tough blow for an economy that was just emerging from the worst years of the Depression.

During this two-week heat wave in early July 1936, record high temperatures were established that have never been equaled. For 11 consecutive days, Quad Citians endured temperatures of more than 100 degrees. Four of those readings were higher than any previous Quad-City temperature.

When you compare the death toll of 72 people to other disasters, there is no other tragedy in Quad-City history that measures up. This death toll, in fact, is higher than the fatalities from all the floods, tornadoes, and blizzards of the last century combined. And the suffering—in the absence of air conditioning, refrigeration, and in some cases electricity—was extreme. In my opinion, this is the greatest weather tragedy in the history of the Quad Cities. And if meteorologist Shipman were still alive, he might be heard to say that hotter breezes might return someday, but it is such a remote possibility that he does not include it in his weather forecast.

During the heat wave, ads for products ranging from early-day air conditioners to soap to prevent body odor proliferated.

Heat Index (Apparent Temperature) Chart

The **Heat Index** (HI) is the temperature the body feels when heat and humidity are combined. The chart below shows the HI that corresponds to the actual air temperature and relative humidity. NOTE: This chart is based upon shady, light wind conditions. **Exposure to direct sunlight can increase the HI by up to 15°F**

Heat Index	General Effect of Heat Index on People in Higher Risk Groups
80 to 89° - Caution	Fatigue possible with prolonged exposure and/or physical activity.
90 to 104° - Extreme Caution	Sunstroke, heat cramps and heat exhaustion possible with prolonged exposure and/or physical activity.
105 to 129° - Danger	Sunstroke, heat cramps and heat exhaustion likely, and heatstroke possible with prolonged exposure and/or physical activity.
130° or higher - Extreme Danger	Heat/sunstroke highly likely with continued exposure.

Relative Humidity (in percent)

Air Temp (in °F)	0	5	10	15	20	25	30	35	40	45	50	55	60	65	70	75	80	85	90	95	100
120	107	111	116	123	130	139	148														
115	103	107	111	115	120	127	135	143	151												
110	99	102	105	108	112	117	123	130	137	143	150										
105	95	97	100	102	105	109	113	118	123	129	135	142	149								
100	91	93	95	97	99	101	104	107	110	115	120	126	132	138	144						
95	87	88	90	91	93	94	96	98	101	104	107	110	114	119	124	130	136				
90	83	84	85	86	87	88	90	91	93	95	96	98	100	102	106	109	113	117	122		
85	78	79	80	81	82	83	84	85	86	87	88	89	90	91	93	95	97	99	102	105	108
80	73	74	75	76	77	77	78	79	79	80	81	81	82	83	85	86	86	87	88	89	91

Dew Point (in °F)

Air Temp (in °F)	60	61	62	63	64	65	66	67	68	69	70	71	72	73	74	75	76	77	78	79	80	81	82	83	84	85
104	110	110	110	110	110	110	111	112	113	114	115	116	117	118	119	121	122	124	125	127	128	130	132	133	136	137
102	108	108	108	108	108	108	109	110	110	111	112	113	114	116	117	118	119	121	122	124	126	127	129	131	133	136
100	106	106	106	106	106	106	106	107	108	109	110	111	112	113	114	115	117	118	119	121	123	124	126	128	129	132
98	103	103	103	103	103	103	104	105	105	106	107	108	109	110	111	113	114	115	117	118	120	121	123	125	127	129
96	101	101	101	101	101	101	101	102	103	104	105	106	107	108	109	110	111	112	114	115	117	118	120	122	124	127
94	98	98	98	98	98	98	99	100	100	101	102	103	104	105	106	107	108	109	111	112	114	115	117	119	122	124
92	96	96	96	96	96	96	97	97	98	99	99	100	101	102	103	104	105	106	108	109	110	112	114	116	119	121
90	94	94	94	94	94	94	94	95	95	96	97	98	98	99	100	101	102	103	105	106	107	109	110	113	116	117
88	88	88	88	89	89	90	90	90	91	92	93	94	95	96	97	98	99	100	101	103	104	106	108	110	112	114
86	86	86	87	87	87	88	88	89	89	90	91	91	92	93	94	95	96	97	98	100	101	102	104	106	108	110
84	84	84	85	85	85	86	86	87	87	88	88	89	90	90	91	92	93	94	95	96	97	98	100	101	103	--
82	82	83	83	83	82	84	84	85	85	86	86	87	87	88	88	89	89	90	91	92	93	94	95	--	--	--
80	80	81	81	81	82	82	82	82	83	83	83	84	84	84	85	85	85	86	86	87	87	--	--	--	--	--

Snow, Wind & Bone-Chilling Cold:
Winter in the Quad Cities

When it comes to snow, I'm one of those oddballs. I am the guy who can't stop smiling when the forecast calls for a foot of wind-blown powder. I stay up all night to see the first flake. I can't wait to shovel the driveway with a beat-up $10 shovel. Give me snow, piles of it, the more the better.

This attitude can get a TV weathercaster into trouble. When I call for snow, many of my viewers cringe and hope for a busted forecast. To keep myself from smiling, I have to think of something sad before I break the news that heavy snow is on the way. This is my dilemma as the bearer of snowy news.

Take it or leave it, during winter in this part of the world, snow will fall. In a typical year, about 3 feet of the white stuff will accumulate. In late October and much of November, the snow is little more than a nuisance, a warm-up for bigger things to come. And, even if it sticks, the chances are it won't be around in a couple of days.

Our first measureable snow in the Quad Cities generally happens within a week or two of Thanksgiving. As we get deeper into December, the possibility of a major storm grows. When the calendar says January, it is time for the mother of all winter storms, the blizzard.

RIGHT: *A March 1966 blizzard that nearly buried telephone poles.*
COURTESY OF NOAA/HISTORIC NWS COLLECTION

FACING PAGE: *A snow-covered dock on the Mississippi River.*
COURTESY OF QUAD-CITY TIMES

Nice fat Charlie Brown flakes—good for eating! COURTESY OF *QUAD-CITY TIMES*

All loaded up with nowhere to go—cars and trucks stranded by a blizzard, west of the Quad Cities.
COURTESY OF *QUAD-CITY TIMES*

The mere mention of the potential for a blizzard spreads alarm through the heart of a city and the media, but even though blizzards are big news events, they are not common. The National Weather Service defines a blizzard as a winter storm with winds of more than 35 miles per hour and visibility that is reduced—for a minimum of three hours—by falling or blowing snow to a quarter of a mile or less. A severe blizzard generates 40-mile-per-hour winds and is accompanied by temperatures of less than 20 degrees. A storm of that caliber comes around one or twice a decade and, in eastern Iowa, it will most likely to occur in January.

The term blizzard may have its roots in the English midlands in the early nineteenth century, when a storm of severe wind and snow was labeled a blizzer. About that same time here in the United States, the word blizzard was used by settlers to describe a cannon shot, a volley of musket fire, or a severe blow. Iowans can be proud that the first time the word blizzard was used in the context of weather was in the Estherville, Iowa, newspaper, the *Vindicator*. It used the word blizzard in an April 1870 issue to describe an exceptional snowstorm that struck Iowa.

When a blizzard pays a visit to Iowa, chances are it was born in Texas. The area from the Texas panhandle east to the Sabine River is considered fertile breeding ground for what meteorologists call the "Texas hooker." This type of snowstorm gets its name from the sharp hook, or left turn, it takes in its path over the Lone Star State. Snowstorms that approach from the south scoop up large amounts of Gulf moisture, which they drop as snow. When a hooker heads from St. Louis, Missouri, to Gary, Indiana, Iowa is in line for a big dump of snow.

The Alberta Clipper is another type of storm that delivers a nasty punch. Named for clipper sailing ships that were once the fastest on the sea, this storm is a speed demon. Once the storm forms on the nose of an arctic air mass that moves south out of Alberta, Canada, it can sail along at 50 miles per hour. Because it moves so quickly, it picks up lit-

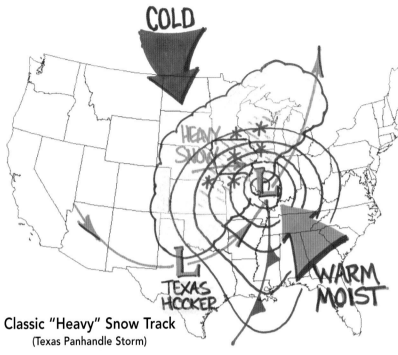

Classic "Heavy" Snow Track
(Texas Panhandle Storm)

Storm track "the big one" is likely to take. COURTESY OF TERRY SWAILS

Snowflakes are actually ice crystals that form in hexagonal patterns on ice and chemical substances floating within a cloud. These crystals eventually clump together to form snowflakes. If the air is dry and cold, the flakes tend to be smaller. If the air is moist and temperatures are near freezing at the ground, snowflakes tend to stick to one another and grow rapidly. This is when those big, fat Charlie Brown snowflakes fall. Even though they seem to be dropping quickly, snowflakes can take as long as two hours to reach the ground from the clouds where they form. Even the biggest flakes fall at a rate of just one mile per hour.

tle moisture, so snowfall is rarely more than a few fluffy inches. As the clipper moves south, however, it generates powerful northwest winds that can blow at speeds of 40 to 60 miles per hour. The winds can cause severe blowing and drifting snow, which create whiteout conditions that lead to road closures. With its sharply falling, often sub-zero temperatures, this storm can be life-threatening for anyone unfortunate enough to get stranded in it.

In any given year, snowfall in this part of the Midwest varies widely and is highly dependent on the track of individual storms. In eastern Iowa, the difference in snowfall between the north and south is very sharp. Yearly snowfall averages range from a high of 50 inches in northeast Iowa to a low of 20 inches in the southeast. In some years, nearly 100 inches have piled up in the north; in other years, less than 10 inches were measured in the south.

The formation of snow is an interesting phenomenon.

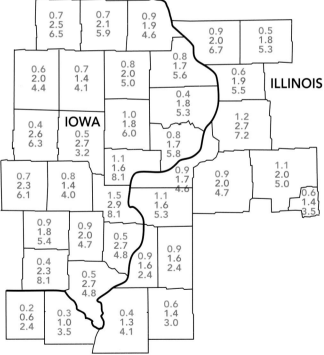

Red: 6 inches or more
Blue: 4 inches or more
Green: 2 inches or more

County-by-County Snowfall — 1995–2004
Annual Average Number of 24-Hour Snowfall Events

106

RECORD SNOWFALLS IN THE QUAD CITIES
(Period of Record starts October 1, 1884)

Seasonal Snowfall (from July to June)

Highest	69.7 inches	1974 to 1975
Lowest	11.1 inches	1901 to 1902

Calendar Year (January to December)

Highest	71.1 inches	1997
Lowest	7.1 inches	1922

Winter (December to February)

Highest	52.9 inches	1978 to 1979
Lowest	6.0 inches	1921 to 1922

Highest Snowfall in One Month

32.0 inches	December 2000

Highest Snowfall from a Storm

18.4 inches	January 12 to 14, 1979

Highest Snowfall in a Day

16.4 inches	January 3, 1971
15.1 inches	January 19, 1995
14.8 inches	January 13, 1979

Greatest Snow Depth

28 inches	January 14 to 19, 1979

MISCELLANEOUS QUAD-CITY RECORDS

Earliest trace of snow	0.1 inch	September 25, 1942
Earliest "big" snow	6.6 inches	October 26, 1967
Average date of first snow	0.4 inch	November 21
Latest first snow		January 4, 1913
Latest "big" snow	7.8 inches	April 17 to 18, 1912
Latest measurable snow	0.3 inch	May 3, 1935
Latest trace of snow		May 22, 1917
Average date of last measurable snow		March 27
Average date of earliest measurable snow	1.5 inches	February 19, 1908

A 1926 manual put out by Good Roads Machinery.
COURTESY OF NOAA/HISTORIC NWS COLLECTION

On January 28, 1887, ranch owner Matt Coleman reported that a snowflake 15 inches wide and 8 inches thick landed on his spread in Fort Keogh, Montana. When asked to give a report for the *Monthly Weather Review* magazine, he said the snowflakes he'd seen were "larger than milk pans." In Bratsk, Siberia, in 1971, another monster flake was observed and measured to be 12 inches by 8 inches. Try catching that on your tongue!

In the Midwest, snowstorms that produce more than 10 to 14 inches of accu-

Weather Folklore

If autumn leaves are slow to fall, prepare for a cold winter.

THE SWAILS SNOWSTORM SCALE

Category	Description	Average Rate of Snowfall	Average Snowfall Accumulation	Minimum Visibility	Winds
1	A **weak storm,** with light snowfall and minimal accumulation. Travel impaired slightly.	Less than 1 inch per hour	Less than 1 inch	1 to 2 miles	0 to 20 mph
2	A **marginal storm,** with a burst of heavier snow that can reduce visibility temporarily. Travel is impaired during the heavy snow burst but is not difficult overall.	Less than 1 inch in 2 hours, with a brief period of 1 inch per hour snowfall	2 to 4 inches	.75 to 1 mile	5 to 25 mph
3	A **moderate storm,** with moderate snowfall during half the storm's duration. Travel at storm's onset is impaired and becomes difficult by the storm's end.	1 inch per hour	4 to 6 inches	.50 to .75 mile	10 to 30 mph
4	A **significant storm,** with moderate snowfall throughout half of the storm and a period of heavy snow also likely. Travel is difficult hours after the storm and very difficult by storm's end. Blowing and drifting of snow.	1 to 1.5 inches per hour	6 to 8 inches	.25 to .50 mile	20 to 45 mph
5	A **major storm,** with moderate snowfall during most of the storm and periods of heavy snowfall. Significant blowing and drifting snow. Blizzard parameters are usually met by the storm. Travel is very difficult at the storm's beginning and nearly impossible at the storm's end.	1 to 2 inches per hour	8 to 12 inches	.10 to .50 mile	25 to 50 mph
6	A **severe storm,** with moderate to heavy snowfall throughout and periods of near-whiteout conditions. Severe blowing and drifting snow, blizzard conditions occur through much of the storm. Travel impossible by the storm's end.	1 to 4 inches per hour	More than 1 foot	0 to .10 mile	30 to 60 mph

mulation are rare. This is because the moisture necessary for such a storm must be transported from the Gulf of Mexico, more than a 1,000 miles away. In other parts of the country that are closer to moisture sources, snow is often measured by the foot, as in mountainous country. In Mount Shasta, California, from February 13 to 19, 1959, a lethal combination of cold temperatures and moist air produced 189 inches of snow, the record amount of snow for a single U.S. snowstorm. Mount Baker Lodge in Washington State holds the record for the most snow during a season—1,140 inches of snow (95 feet) fell during the winter of 1998 to 1999.

Some of the most intense snowfall rates anywhere in the world are generated by lake-effect snow squalls that surge off of Lake Ontario. On January 9, 1976, intense snow squalls in the village of Adams, New York, dumped

Snow fence, properly placed, holds back the drifts from the highway, as shown here.

How to create the perfect snow fence—information from the 1926 Handbook of Snow Removal.
COURTESY OF NOAA/HISTORIC NWS COLLECTION

68 inches of snow in one day, making it the state's largest 24-hour snowfall. Equally impressive was a January 26, 1972 snowburst at Oswego, New York, that produced 27 inches of snow in just 3.5 hours. Do the math: that computes to an incredible snowfall rate of almost 9 inches per hour!

Three out of every four years in the Quad Cities, we can expect a 6-inch snowfall, and, one year out of four, we can expect a 9-inch dump of snow during a single storm. On average, we have about 50 days each winter with at least one inch of snow on the ground and close to 14 days where the snow depth is 6 inches or more. During the winter of 1978 to 1979, the Quad Cities experienced the longest consecutive period—78 days—with at least an inch of snow on the ground and also set our record for snow depth when 28 inches blanketed the ground.

Many people think that it won't snow if the temperature is too cold. This is not true. As long as there is some source of moisture in the air, it can snow at incredibly cold temperatures. What is true is that most of our heavy snowfalls occur when ground temperatures range from 25 to 30 degrees. It's a simple fact of physics—warm air can hold more water vapor and, therefore, produce more snow.

Dreaming of a White Christmas

Many people hope for a white Christmas. As it turns out, the dream of a white Christmas comes true only 40 percent of the time. On December 25, 2000, however, we had a white Christmas worthy of Irving Berlin when 11 inches of snow made it the whitest Christmas on record—and the temperature of 18 below zero made it the coldest! (A white Christmas is defined as a snow-cover of at least 1 inch.)

The Deep Freeze

Many moons ago, when I was a spirited teen, the Swails driveway was more than a parking spot—it was the local

Wind Chill Chart

Temperature (°F)																		
Calm	40	35	30	25	20	15	10	5	0	-5	-10	-15	-20	-25	-30	-35	-40	-45
Wind (mph)																		
5	36	31	25	19	13	7	1	-5	-11	-16	-22	-28	-34	-40	-46	-52	-57	-63
10	34	27	21	15	9	3	-4	-10	-16	-22	-28	-35	-41	-47	-53	-59	-66	-72
15	32	25	19	13	6	0	-7	-13	-19	-26	-32	-39	-45	-51	-58	-64	-71	-77
20	30	24	17	11	4	-2	-9	-15	-22	-29	-35	-42	-48	-55	-61	-68	-74	-81
25	29	23	16	9	3	-4	-11	-17	-24	-31	-37	-44	-51	-58	-64	-71	-78	-84
30	28	22	15	8	1	-5	-12	-19	-26	-33	-39	-46	-53	-60	-67	-73	-80	-87
35	28	21	14	7	0	-7	-14	-21	-27	-34	-41	-48	-55	-62	-69	-76	-82	-89
40	27	20	13	6	-1	-8	-15	-22	-29	-36	-43	-50	-57	-64	-71	-78	-84	-91
45	26	19	12	5	-2	-9	-16	-23	-30	-37	-44	-51	-58	-65	-72	-79	-86	-93
50	26	19	12	4	-3	-10	-17	-24	-31	-38	-45	-52	-60	-67	-74	-81	-88	-95
55	25	18	11	4	-3	-11	-18	-25	-32	-39	-46	-54	-61	-68	-75	-82	-89	-97
60	25	17	10	3	-4	-11	-19	-26	-33	-40	-48	-55	-62	-69	-76	-84	-91	-98

Frostbite Times: 30 minutes | 10 minutes | 5 minutes

$$\text{Wind Chill (°F)} = 35.74 + 0.6215T - 35.75(V^{0.16}) + 0.4275T(V^{0.16})$$

Where, T=Air Temperature (°F) V=Wind Speed (mph) Effective 11/01/01

COURTESY OF NATIONAL WEATHER SERVICE, DULUTH, MINNESOTA

neighborhood basketball court. Day or night, the sound of a thumping Spaulding could be heard, echoing up and down the street.

In the winter, we shoveled the snow or scraped the ice off the court. Many winters, we shot jumpers and played Horse when the temperatures were in the teens and snow lined the court. Inevitably, we would strip off our coats and gloves until my mother would remind us to put them back on.

When I think about that now, I wonder how I could have tolerated such cold. But cold is relative. What seems tolerable as a kid seems unimaginable as an adult. What is an average winter day in Alaska is a frigid one in Florida.

So how do we meteorologists determine what is or isn't cold? With centuries' worth of statistics, we can form a pret-ty good idea of what is "normal" for any time of year. When winter temperatures in the Midwest dip below that "normal" threshold by more than 15 degrees, we are in the refrigerator. When that number drops to more than 30 degrees below normal, we are into record-breaking territory with lows ranging from 15 to 25 degrees below zero. That, in my book, is the good old-fashioned cold that freezes your nostrils and make the tears ice up the corners of your eyes.

Wind Chill

With a little wind, things get a whole lot uglier. This added chill is the result of what meteorologists call evaporative cooling. This is similar to the chill that you feel when you climb out of the shower or swimming pool and the water evaporates from your body. Sweat has a similar effect. We all have a layer of perspiration on the top of our skin that evaporates when exposed to wind. The stronger the wind blows against our skin, the greater the evaporation from our face, hands, and toes and the colder we feel. This heat-stealing effect is what we call the wind chill factor.

The term "wind chill" was coined by explorer Paul Siple in 1939. During an experiment in Antarctica, Siple measured how long it took a can of water hanging on a pole to chill under different combinations of wind and temperature. His formula for wind chill was completed two years later, but it was not part of the National Weather Service forecasts until the early 1970s.

The original formula, while ground-breaking, had a flaw. It measured winds at 33 feet above the ground, while the majority of us exist at ground level where winds are about one-third weaker. In November 2001, the National Weather Service revised the formula by using wind speeds measured at 5 feet.

While the new formula is more accurate, its results are less dramatic. For example, at 5 degrees with a 30-mile-per-hour wind, the old formula measured wind chill at 40 degrees below zero. The new formula measures wind chill at 19 below zero. Ah, the good old days!

Good at math? Try calculating wind chill using the new formula:

Wind Chill T (wc) = $35.74 + 0.6215T - 35.75 (V^{0.16}) + 0.4275T(V^{0.16})$

In the equation, wc is the wind chill in degrees Fahrenheit, V is the wind speed in miles per hour, and T is the temperature in degrees Fahrenheit.

Minutes to Frostbite

Wind speed at mph

Air Temperature in °F

	10	5	0	-5	-10	-15	-20	-25	-30	-35	-40	-45	-50
5	>2h	>2h	>2h	>2h	31	22	17	14	12	11	9	8	7
10	>2h	>2h	>2h	28	19	15	12	10	9	7	7	6	5
15	>2h	>2h	33	20	15	12	9	8	7	6	5	4	4
20	>2h	>2h	23	16	12	9	8	8	6	5	4	4	3
25	>2h	42	19	13	10	8	7	6	5	4	4	3	3
30	>2h	28	16	12	9	7	6	6	4	4	3	3	2
35	>2h	23	14	10	8	6	5	4	4	3	3	2	2
40	>2h	20	13	19	7	6	5	4	3	3	2	2	2
45	>2h	18	12	8	7	5	4	4	3	3	2	2	2
50	>2h	16	11	8	6	5	4	3	3	2	2	2	2

COURTESY OF NATIONAL WEATHER SERVICE, DULUTH, MINNESOTA

Baby, It's Cold Outside

Have you noticed that once snow is on the ground, the temperatures become colder? This is because of an effect meteorologists call albedo. The albedo is a term that refers to the amount of radiation that reflects off of a surface. Snow, for example, has a much higher albedo than the bare ground. The sun's radiation is reflected by the snow, instead of being absorbed, which means that we will be 5 to 15 degrees colder when snow is on the ground.

Take Your Cold Medicine

Many of us have heard that if you don't dress for the cold, you will catch a cold or flu. Contrary to popular belief, the only thing cold weather does is drive us inside. Once we are inside, we are closer together in a warm environment where viruses can thrive. Once a virus is active, we simply pass it along through human contact.

The graph below shows how many minutes it would take for frostbite to occur at a given temperature and wind speed.

In the Quad Cities, as elsewhere, our worst cold waves involve a stiff breeze. The nastiest of the bunch occurred on Christmas Eve, 1983, when an intense arctic front deepened the cold and kicked up the winds. The average temperature on Christmas Eve day hovered at 15 degrees below zero—the coldest December day in Quad-City history. The punishing winds, screeching at over 40 miles per hour, produced a record-breaking wind chill of 72 degrees below zero based on the old wind chill formula that was still in use. In Sioux City, Iowa, the chill was an astounding 81 degrees below zero!

Manitoba Maulers and Saskatchewan Screamers: Cold Fronts

Similar to snowstorms, cold waves originate in other places. The polar regions of Canada, Alaska and Siberia are ideal for generating bone-chilling air masses. The cold, dense air up north is always on the move and, in early winter, becomes an issue when it forms an Arctic front and heads south.

For years, forecasters have noticed that there is a strong connection between Midwest temperatures and the weather in Alaska. When it is bitterly cold in the Yukon, it is often unseasonably mild in the whole Midwest, along with the Quad Cities. However, when it turns mild in Alaska, the frigid polar air has found a new home, and that new home is often in the Midwest. This thermal connection is called the Alaskan Pipeline, and it's worth keeping an eye on, especially in January.

Forecasters have a number of terms for arctic fronts based on where they originate. The Siberian Express refers to icy air that crosses the North Pole and roars into the United States. Manitoba Maulers and Saskatchewan Screamers pour out of the plains of Canada to ravage the Midwest. Any weathercaster worth his or her salt can't resist these metaphors to describe a frigid arctic blast.

In the Quad Cities, we usually get at least one really good shot of arctic air per year. Some winters that number is substantially higher. Most of our cold waves occur during a six-week period that begins in late December and ends in early February. It may surprise you to know that in the Quad Cities, we get an average of 15 days a winter with sub-zero temperatures. It may horrify you to learn that we had as many as 43 zero-degree days back in the winter of 1977 to 1978. Our coldest winter overall actually occurred one year later when we had an average winter tem-

Weather Folklore

If you have a white Christmas, you'll have a brown Easter.

I have found this to be true 80 percent of the time.
—Terry Swails

WINTER TEMPERATURES OF ZERO OR BELOW-ZERO IN THE QUAD CITIES		
0 Degrees or Less	Duration	Date
Greatest Number of Days	43	Winter 1977 to 1978
Least Number of Days	1	Winter 1881 to 1882, 1877 to 1878, 1905 to 1906, 1930 to 1931
Most Consecutive Days	16	January 27 to February 11, 1895
	15	January 15 to 28, 1940
Average Number of Days	17.3	
125 consecutive hours at or below zero from		January 30 to February 4, 1996
Coldest Average Winter Temperature	14.1	1978 to 1979

COURTESY OF NATIONAL WEATHER SERVICE, DAVENPORT, IOWA

perature of just 14.1 degrees. Cabin fever was running rampant both of those winters!

It's bad enough to have a single night of below-zero cold. It's downright brutal when it lasts several consecutive nights. Expand that to two weeks and you have the most prolonged stretch of below-zero cold in the history of the Quad Cities. From January 27 to February 11, 1895, the low temperature stayed at zero or below for 16 consecutive days. Ouch!

Taking that a step further, in 1996 a massive arctic outbreak sent the temperature below zero on January 30 and it never climbed above that mark until February 4. The 125 consecutive hours of below-zero temperatures is the record for sub-zero cold. During that outbreak, the mercury plunged to 28 below on the morning of February 3, the all-time coldest temperature ever measured in the Quad Cities. In other parts of Iowa, readings plummeted as low as 44 below.

Arctic fronts also produce dramatic temperature swings. On January 18, 1996, the temperature dropped from 56 degrees in the morning to 1 degree below zero at midnight. This 57-degree

NATIONAL WEATHER SERVICE WATCHES, WARNINGS AND ADVISORIES	
Events	Characteristics
High Wind Watch	Issued only if high wind warning criteria are likely to be met in 12 to 48 hours.
High Wind Warning	Sustained winds of at least 40 miles per hour for one hour or more, or gusts to 58 miles per hour or more.
Air Stagnation Advisory	Issued if Minnesota Pollution Control Agency declares an air pollution episode.
Civil Emergency Message	Any emergency declared by city, county or state emergency management. Most often associated with nuclear or other hazardous materials. Issued at the request of emergency management.
Wind Advisory	Sustained winds of 30 to 39 miles per hour lasting for at least one hour.
Dense Fog Advisory	Widespread visibilities less than .25 mile.
Freezing Fog Advisory	Very light ice accumulation from freezing fog.
Frost Advisory	Formation of thin ice crystals on ground or other surfaces. Usually occurs with light wind. Frost may occur even though standard observing level temperatures (5 feet above ground) are in the mid to upper 30s. Issued only during growing season.

COURTESY OF QUAD-CITY TIMES

112

NATIONAL WEATHER SERVICE WATCHES, WARNINGS AND ADVISORIES

Cold Season Events	Characteristics
Winter Storm Watch	Watches are issued before warnings. Watches are normally valid for possible events occurring 12 to 48 hours, but may be extended 6 more hours if uncertainty still exists.
Winter Storm Warning	Heavy snow, 6 inches or more within 12 hours, or 8 inches within 24 hours. Warnings may also be issued for smaller amounts if there is significant blowing snow, low wind chills, sleet or freezing rain.
Heavy Snow Warning	Heavy snow, 6 inches or more within 12 hours, or 8 inches within 24 hours. No other significant winter events such as blowing snow, low wind chills, sleet, freezing rain expected.
Blizzard Warning (Ground Blizzard Warning)	Sustained winds or frequent gusts greater than or equal to 35 miles per hour and falling or blowing snow with visibility less than a quarter-mile for greater than three hours. Ground blizzard only deals with blowing snow.
Ice Storm Warning	Ice accumulations of at least .25 inch expected.
Wind Chill Warning	Widespread wind chills of 35 degrees below zero or lower with winds greater than 10 miles per hour. In some parts of southern Minnesota, the threshold may be 30 below.
Winter Weather Advisory	Normally issued for weather that may cause inconvenience or require caution, and is occurring, imminent, or very likely. Issued for up to 6 inches of snow, freezing rain/drizzle, sleet, blowing snow, or a combination of weather elements. Typically issued for the upcoming 12-hour period, but sometimes extends into the second 12-hour period.
Wind Chill Watch	Typically issued when wind chill values reach 35 degrees below zero with a 10 mile-per-hour wind possible 12 to 48 hours in the future.
Wind Chill Advisory	Widespread wind chills of 25 degrees below zero or lower with winds at least 10 miles per hour. In some parts of southern Minnesota, the threshold may be 20 below.
Freeze Watch	Typically issued when low temperatures of 32 degrees or colder will be possible in the next 12 to 48 hours, during the growing season.
Freeze Warning	Low temperatures are expected to be 32 degrees or colder during the growing season.

COURTESY OF *QUAD-CITY TIMES*

fall was the greatest 24-hour temperature change in Quad-City history. It's tough to dress for a day like that!

While wind and cold can harm people and animals, people also ask me about their effects on inanimate objects, such as cars. The good news is that a car does not sweat or have feelings, so it could care less about a 60-below-zero wind chill. No matter how hard the wind blows, your car or any other inanimate object can never get colder than the actual temperature. When you winterize your vehicle, you just need to make sure the antifreeze is

Winter Travel

Winter weather travel can be extremely treacherous. A little preparation and common sense will go a long way toward keeping you safe. The most important thing you can do when the weather gets rough is to think twice before hitting the road. A blizzard or strong winter storm can turn a quick trip into a life-threatening situation.

If you have to travel, here are some safety suggestions:
- Keep your gas tank near full to avoid getting ice in the fuel lines and tank.
- Check your windshield wipers and keep your washer fluid full.
- Carry extra weight (such as sandbags) in the trunk of your car or bed of your truck, particularly in vehicles with rear-wheel drive.
- Take a cellular phone with charged batteries, a CB or ham radio.
- Carry a winter storm survival kit that includes sleeping bags; booster cables; first-aid kit; flashlight with extra batteries; high-calorie, non-perishable foods such as candy bars; knife; sand or cat litter; shovel; windshield scraper and brush; warm clothing, and water.
- Tell someone where you are going, what route you will be taking, and when you expect to arrive.
- Check forecasts and road conditions before departure.
- Bring an NOAA Weather Radio to keep track of changing conditions.

When you are on the road, here are some suggestions for driving safely:
- Turn on your headlights and stop occasionally to brush off snow.
- Slow down.
- Turn, brake, and accelerate gradually.
- Leave plenty of room between you and the other vehicles.
- Be careful on ramps, bridges, and overpasses.
- If you encounter a snowplow, give it room to pass. Use extreme caution if you attempt to pass it—the blade on a snowplow extends several feet in the front and to the side.

If you become stranded, here are some tips:
- Don't panic. Stay in your vehicle and don't try to walk to safety.
- If you have a cellular phone, CB, or ham radio, call for help.
- Attach a cloth to your car's antenna or window to indicate that you need help.
- Turn on the dome light and flashers to make your car more noticeable.
- Periodically run the engine and heater, but make sure you open your window a crack and make sure the exhaust pipe is free of snow.
- Watch for signs of hypothermia and frostbite.

Snow Watches, Advisories & Warnings

The National Weather Service issues a **winter storm watch** when a significant winter storm may affect an area in 12 to 48 hours. A watch will often be issued when meteorologists are still uncertain about the path and strength of a developing winter storm. A watch means it's time to get ready for a storm.

The National Weather Service issues a **winter storm advisory** when winter weather might be inconvenient but not life-threatening. These conditions include snow 3 to 5 inches deep, less than a quarter-inch of ice, blowing snow that reduces visibility and wind chills from 30 to 50 degrees below zero. An advisory means that conditions are hazardous and people need to use caution when driving.

The weather service will upgrade a watch to a **winter storm warning** when the nature and location of the weather event becomes apparent. When a watch is issued for an area, it means that winter weather is imminent or occurring and it is time to prepare for it. Warnings are issued for snowfalls of 6 inches or more in 12 hours, or 8 inches in 24 hours, dangerous ice more than a quarter-inch thick on the roads or wind chills of 50 degrees below zero or lower. A warning means that the weather could be life-threatening. Travel is not recommended.

Finally, the weather service issues a **blizzard warning** when there are winds of 35 miles per hour, considerable amounts of falling or blowing snow for at least three hours, and visibility is reduced to less than a quarter-mile. A blizzard is the most dangerous winter storm because fierce winds and snow reduce visibility to near zero and create wind chills well below zero. Do not travel.

PHOTO COURTESY OF *QUAD-CITY TIMES*

potent enough to protect you from the lowest potential air temperature—but you don't need to worry about the wind chill temperature.

Slip-Sliding Away:
The Perils of Freezing Rain

Once in a while, warm, moist air surges into the Midwest and spreads over a shallow layer of cold air at ground level. When the warm air layer is thick enough and ground temperatures are below freezing, the resulting inversion makes conditions ripe for freezing rain or ice storms.

Of all the winter weather hazards in the Midwest, ice storms are by far the most dreaded and cause the most expensive damage. Along with horrific travel and walking conditions, heavy icing ruthlessly coats trees and power lines. Snapped branches or power lines can cause massive power outages, especially if strong winds are present.

Along with the Northeast, the Pacific Northwest, and the Mid-Atlanic regions, the Midwest is frequented by ice storms. The interior Northeast has far more ice storms than any other region.

In the Quad Cities, freezing rain can occur any time between November and April, but January and February are the most common months for ice storms. Because of its southern location, which keeps it closer to the warmer air, Burlington, Iowa, sees more hours of freezing rain than the Quad Cities.

Most winters, the Quad-City area has about four days where freezing rain makes us slip and slide. Some years, however, we don't have any freezing rain; other years we have as many as ten days. In the years 1948 to 2000, the highest number of freezing rain days was 14 in 1949.

When freezing rain or freezing drizzle is expected to occur and possibly cover surfaces, resulting in hazardous travel, the National Weather Service will issue a winter weather advisory. When ice accumulations are expected to reach a quarter of an inch or more, the weather service issues an ice storm warning.

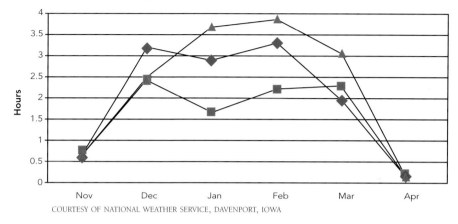

Average Hours of Freezing Rain per Month
Burlington, Iowa; Moline, Illinois; Dubuque, Iowa

COURTESY OF NATIONAL WEATHER SERVICE, DAVENPORT, IOWA

115

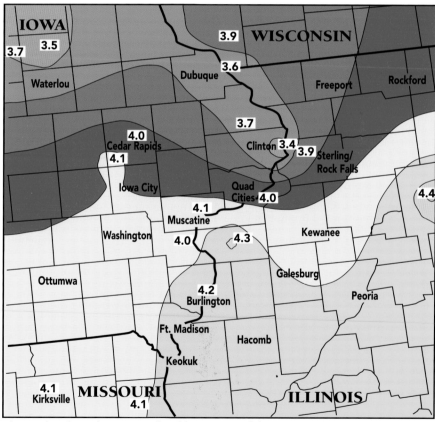

The map above depicts the average number of days per year with freezing rain. Most locations average 4 days per year, but this varies from 0 to 10 days annually. COURTESY OF NATIONAL WEATHER SERVICE, DAVENPORT, IOWA

The Telephone Wire Winter of 1936

ABOVE LEFT: *During the winter of 1936, snow was so high that horse-pulled bobsleds cleared farm fences.* COURTESY OF STATE HISTORICAL SOCIETY OF IOWA

ABOVE RIGHT: *When rural travel was possible, it was often confined to one rutted lane.* COURTESY OF STATE HISTORICAL SOCIETY OF IOWA

RIGHT: *In February 1936, snow buried the locomotives of the Chicago & Northwestern Railroad's Eagle-Grove-Hawarden line.* PHOTO BY RALPH TREMAINE/COURTESY OF STATE HISTORICAL SOCIETY OF IOWA

The winter of 1936 became known as the Telephone Wire Winter. When knees and fenders and fence posts no longer served as adequate comparisons for the mountainous drifts of snow that swallowed trains, telephone wires came in handy. The snow was so high that, according to a January 24, 1936 article by Jim Pollack in the *Des Moines Register,* Lloyd Keller walked from Clarksville to his job at the Iowa State Teachers College in Cedar Falls among drifts so high "he touched telephone wire." At Randalia, Miss Dorothy Moore drove a team of mules from her home to the mailbox a mile away and the snowbanks were so high the ears of the animals touched telephone wires.

Officially, the winter of 1936 was the second coldest and the fourth snowiest winter in Iowa history. Unofficially, the 36-day stretch from January 18 through February 22 was the Super Bowl of Iowa winters. During that period, the statewide temperature averaged 2.4 degrees below zero, and one blizzard was followed by another.

Paul Waite, Iowa's climatologist from 1959 to 1973, said the winter of 1936 was "the worst stretch of winter weather that [Iowa] ever had for both cold and for snow" since the state began keeping records in 1819.

Crunchy Snow

When you are walking on a bitterly cold night, have you noticed how the snow crunches under your feet? What you are actually hearing are the death throes of ice crystals. Snowflakes, which are composed of ice crystals, contain a small amount of air. When snow falls to the ground, the air is trapped inside the layer of snow. When the snow is stepped on, it gets compressed and the air is pushed out of the snow. The sound you hear is the crunching of ice crystals as they break. Try it with ice cubes. They make a crunching sound when they are broken.

Pollack, in his article, described that winter as a 35-day "climatological torture rack." He went on to cite some of the statistics:

Up in Park Lake, the daily high temperature would carry a minus sign on 18 of those days. Down in Red Oak, 43.9 inches of snow fell in January. Heavy snow covered the whole state January 17, then more snow and howling winds made for blizzard conditions on the January 22. Another blizzard hit February 8.

For days on end, Pollack said, Iowans experienced full-scale whiteouts. "Strong winds drifted the snow which fell almost every day during the first 17 days of February," said Charles Reed, the state climatologist from 1918 to 1944. "[This] greatly impeded all transportation."

Pollack reported that the snow was so high that caterpillar tractors with plows couldn't plow the roads, so they were replaced by farmers with shovels. Near Sac City, Pollack's 1986 report stated, "a county road crew and nearby farmers struggled six and a half hours to clear a path so Dr. F. C. Jackson could get to farmer Norman Elam, suffering from pneumonia and mumps 11 miles away." The Works Progress Administration sent 1,600 men out to shovel Des Moines' streets. The city appreciated the help—by February 1, the city's $8,666 winter snow removal budget had been exceeded

by $8,8579. A cream dealer in Marshalltown, Pollack reported, had to deliver his cream by sled, but several days later, "his horse had to be shoveled out of the drifts."

The snow had also stopped the movement of trains. Pollack reported that the 61 passengers cleaned out local grocery stores when the passenger train from Minneapolis to Des Moines met a snowdrift it couldn't bash through near McCallsburg. On another snowbound train near Kellogg, Pollack reported, "passengers amused themselves Sunday night by watching a farm house window, where the farmer undressed to his underwear and went to bed."

Temperatures were as nasty as the snow was deep. Pollack reported that in Lake Park, Iowa, the daily highs for one mid-February stretch ranged from 3 degrees below zero to 15 degrees below zero.

The ice was 42 inches thick on the Iowa River at Iowa Falls, Pollack reported. He quoted a February 1936 report in which state climatologist Reed said the Floyd River was "frozen to the bottom." Water mains froze, leaving 22 Centerville homes without water on February 8. The following day, 500 water meters froze in Des Moines.

117

Plows and tractors couldn't buck the snow and were replaced by farmers with shovels during Iowa's most prolonged spell of cold and snow, January 18 to February 22, 1936. COURTESY OF THE *DES MOINES REGISTER*

All-aboard! The Chicago & North Western's drift-busting attempts to clear the line.
PHOTO BY RALPH TREMAINE/COURTESY OF STATE HISTORICAL SOCIETY OF IOWA

The postal service couldn't deliver mail. The farmers couldn't deliver enough milk and eggs to supply the cities. But by far the most ominous news was that in early February, Iowans were running low on coal. Coal was the primary source of heat in 1936. With high snow drifts, suppliers couldn't find passable routes from the mines to furnaces.

On top of that, Pollack's report stated, Iowa Governor Clyde Herring and the miners' union were not seeing eye-to-eye on the need for the emergency production of coal. Herring ordered the state to share its snow removal equipment with the counties as soon as primary roads were open. "We don't give a damn about the legality," the governor said. "We'll put the snow back if it isn't legal. But these roads must be opened, especially those leading to coal mines."

Herring also wanted the miners to work more hours. Eventually he got his way, after the mayors of Storm Lake and Humboldt, Iowa, called on the governor to take over the mines. The governor even went as far as to say that a mayor had the right to seize coal being shipped through town if an emergency existed.

As supplies dwindled, public buildings were kept open and heated to provide shelter for families that had ran out of fuel. Schools and churches were closed. Anyone who tried to buy coal in Storm Lake was investigated by a city committee to determine their actual need before they were allowed to get a purchase permit. In Radcliffe, families pooled their coal resources and lived together during the shortage, and farmers brought in loads of corncobs and wood to use in place of the coal. Folks in Atlantic burned picket fences and old furniture. Some thief broke into the West Branch home of an elderly, infirm couple—he was 87 and blind; she was 87, and an invalid—and stole their coal.

The Telephone Wire Winter of 1936 had a different effect on different occupations. Gasoline sales dropped at the filling stations, but people were storing more cars there. Telephone operators had hours of extra work after the storms—after the February 8, 1936 blizzard, the Northwestern Bell office in Des Moines put 48 operators to work instead of the usual 23, just to handle all the talk about temperature, coal orders, and weather.

Newspaper deliveries were also a challenge. Trucks loaded with the Sunday *Des Moines Register* left downtown Des Moines at 5 P.M. on February 9, 1936, and by 10 P.M. that evening, they had traveled only 15 miles to Mitchellville. Don Groves, who was the Humboldt County extension agent at that time and later treasurer of the Iowa Farm Bureau, said that it seemed as if he spent all his time "setting up meetings and then canceling them."

The only bright spots in the news were the stories of generosity that poured in, such as that of William Van Horn of Jefferson, who, on a day when it was 17 degrees below zero, hitched four German shepherd dogs to a homemade sled and took 150 pounds of food to

Weather Folklore

It will be a cold, snowy winter if squirrels accumulate huge stores of nuts, beavers build larger lodges than usual, bears and horses develop thicker coats early in the season, and the breastbone of a cooked turkey is dark purple.

75 truckers and miners marooned in a mine 10 miles from town.

Neighbors visited neighbors just to get through the long, snowy evenings. Pollack told the story of one man, Bob Lounsberry of McCallsburg who remembered a snowy evening when the neighbors made a surprise visit. He said his parents were just remarking "'Well, it looks like the Andersons won't be here to play 500 tonight,' but pretty soon we heard the jingle of sleigh bells, and there they came." Nature lovers and sportsmen around the state hauled grain out into the countryside for the birds.

"We were more self-sufficient," recalled Merle Lange, who was the state veterinarian at the time. "We had our sources of food and fuel out on the farm. And when we did need things from town, you'd hitch up a team and bobsled and all the neighbors would join in and go together."

In the Quad Cities, the winter of 1935 to 1936 started off with moderate winter tem-peratures of 35 degrees, and a 47-degree thaw on January 14. A major change was about to send the mercury on a month-long freefall.

On January 16, 1936, snow began to fall. Over a three-day period, 5 inches of snow fell. Then, temperatures plummeted in Prince Albert, Saskatchewan, to 42 degrees below zero. The arctic air blazed south, arriving in the Quad Cities on January 19, 1936, when the mercury registered 1 degree below zero, heralding a 28-day siege of sub-zero cold. The reading of 22 degrees below zero on January 22, 1936 set a new record for the date and was accompanied by stinging winds of more than 30 miles per hour. If wind chill had been calculated at that time, it would have been approximately 60 degrees below zero.

For the rest of January 1936, the snow was minimal, but the relentless northwest wind blew the 9 inches on the ground into significant drifts. Daily temperatures rarely rose above the single digits.

Weather Folklore

If the groundhog sees his shadow, there will be six more weeks of winter. If he doesn't see his shadow, it will be mild.

I have found this to be true 47 percent of the time.

—TERRY SWAILS

EDITORIAL CARTOON FROM THE FEBRUARY 17, 1936 *DAVENPORT DEMOCRAT*

The snow drifts of 1936 made car travel difficult. STATE HISTORICAL SOCIETY OF IOWA

On February 1, 1936, when the ground hog saw his shadow, it foretold what everyone already knew: winter wasn't going anywhere soon. Three more inches of the white stuff fell on February 3 and 4, followed by a low of 15 below. Aside from causing all sorts of problems ranging from frostbite to frozen pipes, the gusty winds closed schools as well as roads that were the supply routes for coal.

On the weather charts, the cold front was nothing more than a line with a series of arrows. For those who felt it pass, however, it was a vicious and powerful. Hopes rose on Saturday, February 8, when the temperatures stood at 32 degrees, the warmest in two weeks. Then, suddenly, flags snapped to attention as the wind veered to the northwest and began to tear across the landscape at 35 miles per hour. The mercury plummeted from 22 degrees to 10 degrees in 15 minutes. The 3 inches of fresh snow, combined with the 12 inches on the ground produced blinding sheets of snow. In less than an hour, normal life slowed to a crawl for residents of the Quad Cities.

By February 10, the temperature had been below zero for nearly 48 hours. Drifts of up to 14 feet choked highways and railroads. In the blizzard's wake, nothing moved. The February 10, 1936 edition of the *Davenport Daily* reported that the Quad Cities were recovering from "the worst blizzard and traffic tie-up Saturday night and Sunday in the memory of local residents. It resulted in one death and the virtual isolation of this community."

The article went on to describe the effects of the storm on the city:

> All highways leading out of the city were blocked most of the day, all train schedules from the west being canceled. Milk distributors in many cases were unable to make deliveries due to their inability to receive supplies from producers in the country and were rationing their customers today.

Another article said that Bettendort and Davenport were "isolated communities" until the early evening on Monday when the huge drifts on East River Street near Forest Road could be cleared. The Bettendorf residents had to use the Bettendorf–Moline Bridge to cross "from the Illinois side to Davenport."

Then, just as residents were recovering, another Siberian Express set its eyes on Iowa. On February 14, after a midnight high of 25 degrees, temperatures crashed the next morning to a frigid 2 degrees below zero. Biting winds swept the landscape, and in no time, roads were closed again. Valentine's Day sweethearts across the Quad Cities postponed their dates.

Snowplows were no match for the hard-packed drifts. According to the Iowa State Highway Commission, at least 75 percent of the state's main highways were blocked, and in some areas, they were worse than in the previous blizzard. By February 18, 1936, snow had covered the ground for 61 days. As winter maintained its stranglehold, the citizens' spirits dropped.

Then, on February 23, the siege was over. The temperature soared to 47 degrees. The following day, when it climbed to 55, one resident said it "felt like 80 after the bitter cold of the previous weeks."

Across Iowa, the period from January 18 to February 22, 1936 was the coldest in 117 years. At its peak, 1 to 2 feet of snow covered the state. Drifts 10- to 15-feet deep were common thanks to the winds, which gusted almost daily. More than half of Iowa's 215,000 farms could not operate for seven weeks because of roads blocked by snowdrifts. Farmers struggled to feed and water animals that consumed twice as much feed as usual because of the severe cold.

By February, it was apparent this would be a winter to remember. Already, fuel supplies were severely depleted. In rural areas, travel was often reduced to horse-drawn bobsleds that rode effortlessly over buried fences and roads. Mail service was delayed up to a week. Railroad service to nearly half of Iowa was delayed for days on end, despite modern snowplows and 100-ton locomotives.

Churches and schools were closed and business hours were curtailed to conserve fuel. Public buildings were opened to families with no heat. Milk shortages were common. For fuel, farmers burned their shade trees and corn cobs that sold for 50 cent a bushel. Hospitals throughout the state reported numerous cases of shock and exposure, some resulting in death. Nearly half of the wildlife in the state perished. In most of the state, frost reached depths of 4 to 7 feet below the ground and the ice on the Mississippi River at Davenport was 26 inches thick.

When the last of the snow melted, survivors could manage a smile, knowing they had endured the most severe winter in Iowa history. What they didn't know was that the summer ahead would be just as extreme. That story was the Heat Wave of 1936.

An Iowan dwarfed by walls of snow never seen before or since. PHOTO BY RALPH TREMAINE/COURTESY OF STATE HISTORICAL SOCIETY OF IOWA

Iowa's Perfect Storm: The April Blizzard of 1973

When I was a junior in high school in Iowa City, 14.3 inches of wind-blown snow fell on my hometown on April 8 and 9, 1973, closing school for two days—something unheard of during the month of April! My memories of the whiteout revolve around the warmth leading up to the storm. It was so nice the day prior to the snow, I had camped out with friends at the Coralville Reservoir. Late that night, it turned cold and rainy, I packed up, drove home, and went back to bed. I awoke later to the voice of my mother telling me to look outside, it was snowing. Shocked, we shrugged it off, knowing it wouldn't last long.

When the last flake fell two days later, I had witnessed the blizzard of 1973, one of the most unusual snowstorms in Iowa history. During the dead of winter it would have been considered one of the state's most severe winter storms. But three weeks into spring, the blast of snow, cold, and wind was stunning and unparalleled. A blizzard of such severe magnitude may not occur again for centuries.

In every way, it was Iowa's perfect storm because so many ingredients had to come together in just the right way. Record cold air had to be in place. Vast amounts of moisture had to be available to fall as snow. And finally, an explosive storm would have to deepen and move along a precise track. The odds of winning the lottery are not much better.

Improbable as it was, everything gelled the morning of April 8, 1973. The snow grew heavier as the day progressed, and when the winds reached 65 miles per hour the next morning, a foot of snow and 5-foot drifts turned much of Iowa into a winter wonderland.

In an article that ran in the *Cedar Rapids Gazette* on April 8, 1993, the twentieth anniversary of the storm, state climatologist Harry Hillaker said several things were unusual about the storm. "It was the combination of being so late, the temperatures so low, the wind speeds were high, [not to mention] the amount of snow," he said.

The snow began in eastern Iowa on Sunday morning, April 8, and didn't wind down until around midnight, April 9. By the time the snow had ended, some cities were hit hard and others went virtually unscathed. Cedar Rapids picked up 14.5 inches. Clinton picked up an inch of snow. Maquoketa, about 25 miles away, reported 16.2 inches. Dubuque recorded 19.2 inches. Belle Plaine received the most snow with 20 inches.

Snow removal was difficult because many state and city plows had been dismantled for the season. By the time they were reassembled, it was too late to keep up with the snow and drifts. The heavy, wet nature of the snow made plowing even more difficult. Some rural roads remained closed for days. Snowmen lined the streets. While the Quad Cities escaped with little more than an inch of snow, less than half an hour west, roads were blocked and closed for at least a day because of the severe blowing and drifting snow.

Fourteen people died in Iowa in weather-related deaths as the heavy, wet snow made travel difficult. Livestock and poultry losses amounted to $20 million. "A lot of the livestock was put on the summer range…for the season and this came up," Hillaker said, adding that 200,000 turkeys and more than 100,000 head of cattle were killed by the storm.

IOWA TODAY

JOHNSON   KEOKUK   MUSCATINE   POWESHIEK   WASHINGTON

—20 years ago today—

'The April Snowstorm'

A pair of pedestrians battle the snow in the Northbrook housing area near Council Street NE.

Unusual spring storm took tremendous toll

PHOTOS COURTESY OF CEDAR RAPIDS GAZETTE

January Thaw

While it's true that January is statistically the coldest month of the year, as we know it's not always cold. In some year, there are usually several days when ole Frosty takes a break. That is the period known as "January Thaw." On average, this welcome respite from winter has a track record of occurring annually during the third and fourth weeks of the month.

Smooth Sledding

Snowmobiling is a popular form of winter recreation over the snowfields of the upper Midwest. Today's high-tech sleds can cost well into the thousands. That's a far cry from the original prototype designed by Edgar and Allen Hetteen and David Johnson in 1954. They created the first sled out of an old toboggan, conveyor belts, and car bumpers. After some revisions they took their second machine 1,100 miles across Alaska in 1960. Facing wolves and grizzly bears, they proved sledding wasn't just for kids!

Field of Mirrors

Frozen water is not just a physical phenomena, it's an aesthetic wonder as well. Take snow for example. On a sunny day a fresh field of it will glisten like a thousand tiny mirrows. Yet in reality it's not the snow you see sparkling, intead it's the reflection of the sun. In and of themselves, snowflakes do not emit light but since they reflect the sun's image back toward your eye they act as tiny mirrors. It's these flaky mirrors that produce the dazzling display of sparkles. What a special effect!

— Autumn's Wrath: The Deadly Armistice Day Storm of 1940 —

This was Excelsior Boulevard in Minneapolis after the November 11, 1940 blizzard.
COURTESY OF MINNESOTA HISTORICAL SOCIETY

124

There is something majestic about a raging winter storm. It has no respect for boundaries or wealth; it goes where it wants, when it wants, and it can bring an entire state to a grinding halt. Nearly 65 years ago, a storm of such incredible magnitude not only changed the landscape of the Midwest, it changed the lives of those who lived it. The Armistice Day blizzard that blew through on November 11, 1940, was one of the most severe storms to ever strike the nation's heartland. With its vicious winds and plummeting temperatures, the storm took the lives of more than 160 people, including many duck hunters who were trapped by the storm on the rivers, lakes, and wetlands of the Midwest.

The warm fall of 1940, usually a time of football and

fiery colors, was overshadowed by the escalating war in Europe. When the first frost finally settled on freshly picked fields of corn and beans on November 7, 1940, it was more than a month behind schedule. Nevertheless, the weather warmed right back up. Within days temperatures topped 60 degrees as far north as Minnesota.

When their bedside alarms rang on the morning of November 11, 1940, the early risers in the Quad Cities were pleased to see that, despite the overnight showers, temperatures were balmy for November. A brisk south wind had kept readings in the 50s. It was Armistice Day, (Veterans' Day since 1945), so many businesses and schools were closed. Duck hunters eagerly headed out to the sloughs of the Mississippi. It wasn't often that bird hunters could bag their prey without the need of a heavy winter coat.

As hunters maneuvered duck blinds and decoys into their favorite spots, a storm was rapidly evolving in central Iowa. Bitter cold air that had been bottle necked in Canada charged south into a low pressure center near Des Moines. In the meantime, warm, moist air from the south streamed into the storm's core. As these two air masses merged, a spin formed around the storms, which dropped the air pressure at an alarming rate. When the barometer crashed to 29 inches, it was apparent that this was no ordinary storm. In fact, it was what meteorologists call a "bomb"—a storm that will do incredible things.

In advance of the storm, the temperature in the Quad Cities soared to 58 degrees by 9 A.M. As thousands of ducks filled the morning sky, warm winds blew and the crack of shotguns echoed throughout the river valley. What few hunters realized was that the ducks were moving out for a reason: they were seeking shelter from the severe weather developing to the west.

Lulled by the unseasonable warmth, many hunters were

unprepared for the change in weather. Coats and gloves that were usually standard gear had been left behind in the car, miles from the men who would need them. For some, it was mistake that would cause extreme hardship. For others, it was a mistake that would cost them their lives.

A vicious cold front raced across the Midwest at speeds of more than 40 miles per hour, blowing through eastern Iowa in few hours. Then it descended on the unsuspecting hunters along the Mississippi River. In seconds, a pleasant, southwesterly breeze shifted to the west and blew at gale-force strength. Whistling into bobbing duck blinds, the wind sprayed water over the boats and their occupants. Two-foot waves and whitecaps soon chopped the river's surface and rocked the boats.

Temperatures plummeted 20 degrees. By 11 A.M., it was 38 degrees and falling. By noon the reading had fallen below freezing to 28 degrees. Falling a degree every six minutes, the mercury had dropped 30 degrees in just three hours. Wind chills had slipped to nearly zero degrees and, in many areas, snow flurries joined the spray showers that had soaked hunters to the bone. It was apparent that anyone on the water was in serious trouble.

As the eye of the storm passed over Charles City, in northeast Iowa, the barometric pressure fell to a record-breaking 28.92 inches. Winds of 50 to 60 miles per hour whipped the waters of the Mississippi River into waves of unheard-of heights. In spots, 5-foot breakers churned across the river. On shore, trees and power lines snapped like toothpicks. Boats that tried to buck the waves careened helplessly and were forced onto islands or shorelines. For many, there was nothing to do but wait until help arrived or conditions improved.

As the storm raged on that afternoon, the local authorities began getting an increasing number of calls from family and friends of hunters who had not returned home. Even hunters themselves, fortunate enough to have escaped the river, called in to tell the authorities of others who desperately needed

assistance. Search and rescue teams were organized, but they couldn't do much until the weather broke.

Darkness settled over the river valley and by 6 P.M., the temperature had shrunk to 18 degrees. Ice coated the clothes of stranded hunters, many of whom had abandoned boats, guns, and decoys to seek shelter. Some of these items were also used as kindling and burned for the precious warmth the hunters needed to survive. As the night progressed, shivering hunters began hitting themselves so they could stay awake—falling asleep in these conditions would almost certainly result in death. When they were found, some of the hunters' bodies were bruised and battered from fighting the battle to avoid freezing to death. Others huddled with dogs in makeshift shelters, praying for calmer winds and an end to the howling night.

The temperature kept dropping. By 8 A.M. on November 12, it bottomed out at 13 degrees—45 degrees colder than

Hunters Marooned on Islands in River; One Muscatine Man Dead

Two hunters who froze to death during the Armistice Day Storm. PHOTO BY MINNEAPOLIS STAR-JOURNAL/COURTESY OF THE MINNESOTA HISTORICAL SOCIETY

majority had died of hypothermia. One man was found frozen upright in the water, his hands clutching a branch. When rescuers found him, they had to cut the branch from either side of his hand.

In the Quad Cities, Leon Reynolds, age 42, of Muscatine, Iowa, drowned when his boat capsized as he tried to reach the Mississippi's shore, north of Burlington. Louis Tubbs of Muscatine, Iowa, and Ralph Sells of Oquakwa, Iowa, managed to move their boat to an island. Fort Madison hunters Franklin Payne, Lawrence Cross, and Douglas McKimm were found dead in a duck blind the following night. Near Savanna, Illinois, Hugh Baily was collecting driftwood for his stove at home. Caught in the tempest, his boat was swamped and he died alone of hypothermia stranded on an island. Many more had narrow escapes.

The weather was even worse to the northwest of the Quad Cities. As it continued to move on a path from Eau Claire, Wisconsin, north to Lake Superior, the storm's central air pressure fell to 28.66 inches. Stronger than many hurricanes, the storm produced winds that gusted to more than 80 miles per hour. North and west of the storm's center—where temperatures were colder yet—ice and snow pounded parts of Iowa, South Dakota, and Minnesota. Primghar, Iowa, reported 17 inches of snow. The Twin Cities were paralyzed by 16 inches. Collegeville, Minnesota, however, took the brunt of the storm with a 27-inch snowfall.

Thousands were stranded, especially in Minnesota, where the wind and snow produced blizzard conditions. Drifts as high as 20 feet completely buried cars, and rescuers had to use long probes in desperate attempts to find missing people. The wind-driven snow was also rock-hard, which made things even more difficult. State snow removal workers needed rotary plows to carve out one-lane trails, and

Weather Folklore

A heavy November snow will last till April.

the previous morning. Come the dawn, the worst of it was over as the storm swept north into Canada.

Then the grim task of counting the casualties began. The news from search crews was astounding. From Winona, Minnesota, to Burlington, Iowa, as many as 40 hunters perished during the storm. Some drowned, but the

many roads remained closed for days. Across the state, snowbound passenger trains littered the frozen landscape.

In Iowa, the toll on livestock and vegetation was staggering. As many as 1,500 cattle, 2,000 sheep, and more than 200,000 turkeys froze to death or suffocated in snowdrifts during the storm. Thousands of the state's pheasants were killed. Approximately 10 to 15 million bushels of corn were destroyed. Virtually all of Iowa's apple trees were killed or severely damaged. Iowa was a major apple producer and, a year after the storm, the apple crop was down 90 percent because of the million-dollar losses inflicted by the storm.

Telephone and power lines were damaged throughout the Midwest. The hurricane-force winds toppled buildings, barns, and outbuildings in Minnesota, Wisconsin, Illinois, Iowa and Michigan. Huge waves on the Great Lakes sent several ships to watery graves.

By the time the blizzard blew itself out on November 12, 1940, 162 people had lost their lives. The majority died agonizing deaths, freezing on the lakes or rivers or in the wilderness of the upper Midwest. Many were carried home in the positions they froze in, as they waited for help that never came.

The Armistice Day Storm would be a storm of deadly proportions today. But with sophisticated weather technology and improved communications, we would be able to predict the storm well in advance, eliminating the surprise factor that was so deadly in 1940. As a result, the loss of life and property would be substantially less. With its extreme combination of wind, temperature, snow, and speed, the Armistice Day Storm was one of the most severe early winter storms to strike the Midwest.

LEFT: *Digging out cars after the 1940 blizzard.* COURTESY OF THE MINNESOTA HISTORICAL SOCIETY

BELOW: *Rotary plow taking a second cut through snowdrifts in Anoka County, north of Minneapolis.* PHOTO BY LYNN/COURTESY OF THE MINNESOTA HISTORICAL SOCIETY

Mercury to Dip Further; Death Toll 116

The Big Chill: January 1979

In late 1978, when the first winds of winter brought out coats, gloves, and hats, the citizens of the Quad Cities went about their business. Winter, after all, comes and goes. Some winters are short; others are long; but winter always arrives. "You don't need a ruler to measure a winter," a grizzled old timer once told me. "All you need to know is the amount of trouble it causes."

January 1979 was nothing but trouble. A remarkable union of snow and cold forged a landscape in the Quad Cities that rivaled that of the Arctic. The siege began on New Year's Eve, when Quad-City party-goers had to cancel their holiday plans because Mother Nature threw an impromptu party of her own. Snow and wind were the unwelcome guests and they made a mess of what was supposed to be a very special night.

Grab a coat and some hot chocolate, you are about to read the chilling tale of January 1979.

New Year's Day, 1979, opened with a fresh 11.2 inches of snow, 30-mile-per-hour winds, and biting temperatures of 10 degrees below zero, heralding an atmospheric alignment that would hold for weeks to come. For the next few months, a merry-go-round of snow, wind, and cold had forecasters dusting off the record books. The weather service phones never stopped ringing as callers asked the meteorologists to explain what was going on and to predict when it would end.

This story began with the jet stream, the polar jet stream to be exact. A fast-moving ribbon of air that meanders around the globe, the polar jet stream rarely stays in one position for long; instead it regularly buckles and flattens. North of the jet stream, the air is cold; to the south, the temperatures are more moderate. The polar jet stream dwells in Canada during the summer, and only occasionally sends evidence south that it's alive and well.

However, come November, as cold air masses grow stronger by the day, the winds of the polar jet stream intensify as the contrast between warm and cold air steadily increases. With time, the shorter days and less direct rays of sun create pools of air so cold and so vast that their weight forces the jet stream to sink south. Ultimately, its destination is the United States.

To create the frigid conditions of the winter of 1979, the polar jet stream had to be in a position to bring cold air as well as moisture. Because the jet stream separates cold air from warm, it acts as a thermal boundary along which storms develop and move. This is what forecasters call the storm track. It is here that the warm, moist air interacts with cold polar air to form intense low pressure systems. Heavy

SHELL-SHOCKED

Bill Campbell's cartoon of a weary snow shoveler reflected the mood of many Midwesterners during the January 1979 deep freeze. COURTESY OF QUAD-CITY TIMES

snow is likely to develop about 150 miles north of these twirling storms.

In January 1979, the storm track was far enough south of the Quad Cities to make the area bitterly cold, but it was also close enough so that moisture-rich snowstorms were frequent visitors. As a rule, this combination occurs only two or three times a winter, if at all. In 1979, it lasted for weeks.

The snow that covered the ground from the Midwest to the Arctic affected temperatures. The weak rays of the January sun were reflected by the snow, instead of being absorbed by the ground and warming the atmosphere. As a result, temperatures were often 10 to 20 degrees colder than they would have been without the blanket of snow. The snow cover also was so extensive that frigid air masses moving south from the Arctic had little chance to warm up Many times, a pool of air that is bitter cold warms as it plunges south over the bare ground of the Canadian prairies or the upper Midwest. Some winters this warmed air can be the difference between a few sub-zero readings or a season with lots of them. In 1979, the extensive snowpack acted like a freezer for the numerous cold waves that scoured the country. Not only did the arctic air stay frigid, it rode as easily as as a sled into the heart of the nation. Night after night, the mercury dipped below zero.

As if to set a benchmark for the month ahead, the morning of January 2, 1979, was the coldest of the entire winter. Fresh from the Yukon, a dome of high pressure settled over eastern Iowa. The cold, dense air, clear skies, light winds, and deep snow made the mercury plunge to 27 degrees below zero, a place it had never been before—an all-time record. Cars groaned and many refused to start. School, which was supposed to resume after the Christmas break, was postponed for a day. Kids with new sleds and skates were anxious to use them, but they could do little more than look out the window—the thermometer never climbed higher than 5 degrees below zero.

Quad-City Times

28 Pages

Tuesday, Jan. 2, 1979

Quad-Cities, Iowa and Illinois

★ ★ 20 Cents

Numb Q-C Begins To Dig Out

The weather was beginning to get on people's nerves. Not only did the snow refuse to melt, chemicals such as salts that are used to melt the ice were ineffective on the roads.

For Davenport, the issue of snow removal was rapidly becoming a big problem. At a January 2, 1979 city council meeting, Harold Ziffren, a Republican alderman, complained about snow removal on downtown streets. He was quoted as saying, "This downtown is the laughing stock of the whole Quad Cities community. If this ever happens again, I as an individual am going to raise so much hell...."

For the next 11 days, cold was the big weather story. The storm track had temporarily shifted south as a series of

Abandoned cars, such as this one, littered the landscape following the weekend blizzard.
COURTESY OF *QUAD-CITY TIMES*

Shovelling snow several feet deep was challenging and regular work during January 1979.
COURTESY OF *QUAD-CITY TIMES*

Cold Hard Facts

The coldest temperature recorded in the United States was 80 degrees below zero on January 23, 1971 at Prospect Creek, Alaska. Every state in the country has had a subzero temperature except Hawaii, although temperatures there have been close. Mauna Kea, with an elevation of 13,770 feet, was 12 degrees in May 1979. In the lower 48 states, Rogers Pass, Montana, was the site of the nation's coldest temperature of minus 70 degrees on January 20, 1954. The coldest temperature east of the Mississippi was 54 degrees below zero at Danbury, Wisconsin, on January 24, 1922.

arctic high pressure systems bridged the upper Midwest. The good news was that the snow could stop because the gulf moisture was blocked. The bad news was that a steady stream of bitterly cold air was parading across the Quad Cities.

During the first 11 days of the month, the mercury never climbed higher than 15 degrees and there were sub-zero temperatures nightly. The average low over that time was a bone-chilling 14 degrees below zero, 27 degrees below the normal of 13 degrees.

Finally, on January 13, 1979, the long, nasty stretch came to an end following a record low of 17 below zero. The 11 consecutive days of sub-zero temperatures did not set a record: that belonged to the winter of 1895 when temperatures were below zero for 16 consecutive days.

Just as the cold was loosening its grip, local forecasters began receiving data that showed air pressures lowering in the Pacific. Satellites and computer models con-

Weather Folklore

If March comes in like a lion, it will go out like a lamb. If March comes in like a lamb, it will go out like a lion.

This is true only 47 percent of the time.
—TERRY SWAILS

firmed that a storm was developing and was fast approaching the Pacific Northwest. As this swirling mass of clouds and energy rode the jet stream into the Midwest, forecast models indicated a complex interaction of moisture and cold would form a very big storm. The weather charts were ominous; a major blizzard was in the works.

In the meantime, things were heating up in Davenport's city hall, where people were debating over city policy and Mayor Charles Wright's involvement in decisions regarding towing and plowing priorities. Stranded cars still littered many streets, and traffic was often down to one rutted lane, even on streets that had been plowed. Snow removal budgets had already been depleted. According to a Davenport city snow removal plan, only 351 miles of the cities 500 miles of streets were scheduled to be plowed. One more major storm and the city would be paralyzed and in financial trouble.

On January 12, local forecasters took a small sigh of relief. The big storm, while alive and well, was taking a track that would keep the heaviest snows southeast of

When Leaves Begin to Fall

When the frost is on the pumpkin, we wonder if we will have an Indian summer. Indian summer is defined as a prolonged period of mild, dry weather following the first hard freeze. Typically, this occurs in October or November. Some years, when the weather is especially nice, we may see more than one Indian summer. The term originated with the American Indians—possibly the Narragansett Indians—who enjoyed this time of year and said it was a special gift of Cautantowwit, the god of the Southwest.

COURTESY OF *QUAD-CITY TIMES*

the Quad Cities. Even so, the system would be close enough to deposit several inches of unwanted powder. By evening, snow had begun to fall hundreds of miles in advance of the storm. The National Weather Service issued a travelers' advisory for Saturday, January 13. As Quad-City residents went to bed that Friday, the official forecast called for a 90-percent chance of snow with accumulations of 2 to 4 inches—a manageable snowstorm.

The first thing a forecaster learns, however, is that nothing ever goes as planned. All storms have their quirks. What was supposed to be a Saturday storm of nuisance proportions became one of the worst blizzards in the history of the Quad Cities.

Like humans, storms are born and go through similar stages of infancy, maturity, and death. Storms, however, grow stronger and more mature at unpredictable times, depending on heat, moisture, energy, and wind. The addition of just one of these ingredients at just the right time can have explosive results.

The night of January 12, the fuse was rapidly burning for the Quad Cities. A storm in Texas was about to make a turn northeast that would take it into the state of Ohio. All day, computer models predicted the storm would stay the course and remain southeast of the Quad Cities. That evening, how-

Snowplows fight to open a rural stretch of highway. COURTESY OF *QUAD-CITY TIMES*

131

ABOVE: *Hand-printed signs were made to indicate buried fire hydrants.* COURTESY OF *QUAD-CITY TIMES*

BELOW: *Better late than never! A man puts snow chains on his truck in the middle of Davenport's Brady Street.* COURTESY OF *QUAD-CITY TIMES*

ever, weather balloons launched from National Weather Service offices detected arctic air rapidly advancing over the plains of the Dakotas and Nebraska. As the cold air surged south, it drew warm, moist air from the Gulf of Mexico northward. All of these ingredients were going to enter the center of the Texas storm faster than originally thought. This would have huge implications on the Quad Cities, meaning that snow would have to be measured by the feet instead of inches. Forecasters scrambled to get a grip on the storm's track and intensity.

This injection of frigid air into the storm's belly was like throwing gas on a fire. Air pressure fell rapidly, which indicated that the storm was growing stronger. The winds of the jet stream further fanned the fire as they channeled warm, moist air to its eastern flank. With its newfound power, the storm would now take a track that would take it farther north. Thunder storms popped in the warm air of Arkansas and Mississippi. Heavy snow, sleet, and freezing rain began to plague much of the Midwest. And despite the fact that the center of the storm was still 1,000 miles away, heavy snow began to fall on the Quad Cities.

The reports that began to trickle into the National Weather Service confirmed the magnitude of the storm. The National Guard was called out in Oklahoma to help thousands who had lost power or had become stranded in cars. The Iowa State Patrol shut down Interstate 80 in western Iowa and in Kansas. In Chicago, the O'Hare Airport closed at noon—it was only the fifth time the airport had closed since World War II. In North Dakota, a howling wind kept the temperature at Devils Lake at 18 degrees below zero.

At noon on January 13, the eye of the storm peered down on Memphis, Tennessee. Snow fell in sheets over the Midwest. A foot of snow fell in the Quad Cities, four times what the forecasters had predicted, increasing the depth of the snowpack to approximately 2 feet. A snow emergency went into in effect.

Wind was also a problem. The pressure gradient that determines the direction and speed of a storm's winds was tightening as it intensified. The storm was spinning like a freshly pulled top, generating 30- to 40-mile-per-hour northeast winds which were piling up drifts and reducing visibility. Most roads outside the city were impassable

As the conditions worsened, so did the problems. By late afternoon on January 13, 1979, all Quad-City roads were impassable. To the north of the city, several hundred people were stranded on Interstate 80. To find shelter, approximately 200 motorists walked from their cars to the Trout Valley housing development between the Mississippi River and Middle Road. Many more struggled to the Black Hawk

Standard service station at the corner of Interstate 80 and Middle Road.

In the Quad Cities, Interstates 80, 74, and 5 were officially closed by 6:00 P.M. The sheriffs' departments in all of the surrounding counties of Iowa and Illinois reported that the roads were closed because of blowing and drifting snow. Throughout much of Iowa, plows were pulled from the road because the snow was too deep to plow and four-wheel drive vehicles were no match for the drifts. An Iowa state trooper in the Quad Cities was quoted as saying, "Do not, I repeat, do not leave your residence."

The snow was so deep and the roads were so impassable that Iowa law enforcement authorities resorted to snowmobiles to make emergency calls. In Cedar County, sheriff's deputies and volunteers on snowmobiles patrolled Interstate 80, removing stranded motorists. In Davenport, Mayor Charles Wright used ten snowmobiles to handle emergencies. Even so, blizzard conditions prevented Scott County Medical Examiner Dr. R. M. Perkins from reaching a man who had a heart attacked at a truck stop. He died at the truck stop and Scott County deputies were told to bring the body to Davenport when the storm had passed.

The Quad Cities, at this point, resembled a snowy ghost town. Virtually every business, restaurant, and retail outlet was closed. Northpark and Southpark malls had been shut down by noon. Peterson Harned Von Maur closed its doors Saturday for only the third time in 106 years. All reservations at the Dock Restaurant on River Drive were cancelled by late afternoon. And, despite the fact that both teams had set foot in the gym in Iowa City, the college basketball game between Iowa and Indiana was called off because fans couldn't get to the gym. Bars and nightspots sported closed signs. Residents were urged to make only essential telephone calls, as phone circuits were overloaded.

As midnight neared, the snow that had fallen furiously all day relented. As the big storm churned away, it took its

Paralyzed!
Worst Snow In Quad-City History

Quad-City **Times**
178 Pages · Sunday, Jan. 14, 1979 · Quad-Cities, Iowa and Illinois · ☆ 60 Cents

canopy of snow with it. In its wake came an arctic front. Winds of 50 miles per hour whipped the fresh powder into massive drifts, reducing visibility to zero in the open country. To make matters worse, the temperature was free-falling. From a two-week high of 20 degrees on Saturday, the thermometer plunged to a record-breaking

Abandoned cars and trucks shut down Interstate 80 near Le Claire, Iowa. COURTESY OF *QUAD-CITY TIMES*

Shivering

Your body is designed to maintain a 98.6-degree temperature. When your body temperature slips below that level, your nervous system sends signal to your brain to warm things up. To accomplish this, your brain sends a return message through your spinal cord to the nerves throughout your body. In response, your muscles begin to rapidly tighten and loosen, which makes you shiver.

low of 21 degrees below zero on Sunday. Wind chills of 50 below made death a near certainty for anyone exposed to the elements.

Two-year-old Jason Palmer of Des Moines was one of the luckiest survivors of the storm. Iowa State Patrol Trooper Dennis Merritt found Mrs. Palmer, her four children, and her brother and sister-in-law, Ted and Wanda Wessel, and their two children, stranded in their car around 9 P.M.

Deputies on patrol checked the families a short time later, but they would not leave the vehicle. When the car ran out of gasoline around 5 A.M., the families walked less than a quarter-mile to a nearby truck stop. Mrs. Palmer, who was carrying her 11-month-old child, lost track of Jason. An hour and a half later, Jason was found, curled under the wheel of an abandoned truck, just minutes from death. He survived after being treated in the burn unit of the University Hospital in Iowa City for severe frostbite on his feet and hands.

There were other survival stories, too numerous to report. Snowmobiles, however, saved the lives of many. Law enforcement officials used snowmobiles to transport people from cars or homes to warm shelters or hospitals for medical attention. Law enforcement officials and volunteers also used snowmobiles to deliver medicine, gasoline, and heating fuel.

The headlines of the Sunday *Quad-City Times* proclaimed, "Paralyzed! Worst Snow in Quad-City History!" Since the beginning of the storm at noon on Thursday, 18.4 inches of snow had fallen, the most ever from a single storm, burying the previous record of 16.4 inches set January 3, 1971. The official snow depth measured 28 inches at 6 P.M., which is still the all-time record. Pretty big things from a storm that was expected to produce totals on the order of 2 to 4 inches!

By the time the sun rose on Monday, January 15, Illinois Governor James Thompson and Iowa Governor Robert Ray declared the Quad Cities an emergency disaster area. Armored personal carriers from the Iowa National Guard headquarters at Camp Dodge were dispatched to the Quad Cities to help with snow removal. Interstate 80 in Iowa, from Iowa City to Bettendorf, was still only one lane. Stranded motorists jammed the truck stops. At the Golden Finch truck stop near Walcott, Iowa, just west of the Quad Cities, 200 persons ate and slept in booths until roads were passable. Police in the Quad

Quad-City Times — Quad-Cities, Iowa and Illinois — Tuesday, Jan. 16, 1979 — 28 Pages — + 20 Cents

Stranded cars and trucks litter Interstate 80 near LeClaire.

(Times photo by Bill McConnell)

More Snow On Way

Cities began to tag and tow the cars that blocked virtually every street.

As the temperature fell to a record-breaking 24 degrees below zero, city plows struggled to move the immense mounds of snow. Donald Hindman, an administrative assistant in the state's emergency service office, said that it was so cold some of the equipment wasn't starting and the operators had to bring the batteries inside to warm them up. Once the plows were moving, the machinery had mechanical problems because they weren't built to move such large amounts of snow. Even graders with heavy chains on their wheels were getting stuck in the drifts.

Rural areas were especially hard hit. Rotary plows were the best means of breaking a narrow trail through 10- to 15-foot snowdrifts packed hard by arctic winds. Most of those, however, were being used to clear the interstates or major highways. National Guard helicopters were called into service to assist with medical emergencies and to ferry supplies to farmers who were isolated by lanes and roads filled to the fence tops with snow. Food was dropped for thousands of

Quad-City Times

56 Pages

Wednesday, Jan. 17, 1979

Quad-Cities, Iowa and Illinois

★ 20 Cents

Fire And Ice
Big Snow Skips Q-C; New Storm Threatens
$22.5 Million!
Last Weekend's Snowstorm Picked The Quad-Cities' Pocket

farm animals that hadn't been fed in days. Other farmers who were unable to feed and water stranded livestock purchased snow blowers and snowmobiles so they could complete their chores.

At hardware stores and farm service dealers, business was brisk. The sale of snow shovels, salt, tires, chains, snowmobiles, and four-wheel drive vehicles was soaring. Many retailers ran out of snow shovels. Jerry Stahle of K & K Hardware estimated his store had sold 1,500 snow shovels and 400 to 500 scoops or sand shovels since the snow kicked in. Food and dairy products were in short supply as the people of the Quad Cities swarmed the grocery stores in the blizzard's wake.

There were two major issues affecting the lives of most of the people of the Quad Cities: the prospect of more snow, and the removal of what had already fallen. Others were more concerned about the weight of the snow as roofs on homes, barns, and businesses collapsed. Near Fulton, Illinois, the chicken house of Glen Bechtel came crashing down, killing 32,000 chickens. "They're still lying under the roof," Bechtel said in the *Quad-City Times*. "I'll have to bury them if I can find a backhoe big enough."

As Bechtel contemplated his dead chickens, meteorologists at the National Weather Service issued a winter storm

Moisture and Cold

What do static electricity, itchy skin, and dry throats all have in common? The answer is low humidity. As air temperatures drop, so does the ability of air to hold moisture. Even though the humidity of the cold air outside is 80 percent, if you bring that same air inside and heat it to 72 degrees, the humidity drops to as low as 2 percent. Compare that to the humidity of the Sahara Desert—and the desert looks downright moist at 25 percent! To feel warmer and free of static electricity, use a humidifier.

135

The White Cyclone of 1885

For the early settlers in what is now the Quad Cities, the winter of 1885 was a test of resolve and survival, especially when it came to the deadly storm dubbed the White Cyclone. The big storm itself was actually preceded by an intense four-hour snowfall on February 7, 1885 that added 8 inches to the existing snowcover. There was, however, no wind or temperature drop.

Snow fell again on the afternoon of February 8, 1885, but this time it was accompanied by falling temperatures and gale-force winds. The storm raged for 36 hours, until the morning of February 10, when the mercury dropped to nearly 30 degrees below zero. Trains stopped; post offices and businesses closed.

Then, just when it seemed that the storm had begun to play itself out, the winds picked up again and, in no time, the blizzard roared back to life. Temperatures hovered near 20 below zero, and whiteout conditions threatened the life of anyone unfortunate enough to be caught in the northwesterly gales.

Over the years, the legend of the storm continued to grow and many a night, survivors told of what it was like to be caught in the White Cyclone of 1885.

The Demon of December: The Blizzard of 1987

Thunder and lightning don't often accompany a snowstorm, but all three were part of a violent late-fall storm that roared through the Quad Cities area on December 15, 1987.

The day before, the storm churned out of the Southwest, rapidly deepening as it targeted the upper Midwest. Forecasters predicted days in advance that it would be a big one for the Quad Cities, but people were still shocked at the severity of the wind and snow.

At the height of the storm, thunder and lightning accompanied winds of 50 miles per hour and snow that fell at the rate of 2 inches per hour. Whiteout conditions prevailed for several hours, and severe drifting made most roads impassable. The total snowfall of 11.4 inches was the greatest snowfall on record for a 24-hour period in December. The weather seemed to ensure a white Christmas, until it warmed up to 45 degrees on Christmas Eve—leaving Santa Claus less than an inch on which to land his sleigh!

watch because more heavy snow was likely. A low pressure center in the Rockies was set to swing east—bringing snow back to the area by Tuesday night. A spokesman for Iowa Governor Robert Ray said simply, "Another snow would be disastrous."

This was especially true in the Quad Cities. Citizens were angrily blasting the city, especially Mayor Charles Wright and Public Service Director Robert Saint Clair, about the slow progress of snow removal. Besieged by endless complaints about snow removal, a supervisor in the county roads department announced his plans to step down because he could no longer tolerate the abusive complaints from irate citizens. County Engineer Elmer Clayton said complaining citizens had been calling him at all hours of day with complaints, some using bad language. "You'd like to think you can sleep those few hours you're at home," Clayton said. "But once your name becomes a household word, they know the person they want to chew out."

Thieves took advantage of the situation. While police officers were occupied by traffic violations, the number of burglaries and petty crimes jumped in Davenport and Moline. "They've got the advantage," said Caption Charles Borgstad of the Davenport police. "Squad cars can't get around as good when the weather is like this."

The police also received more disturbance calls. "People have been cooped up so long they're starting to get cabin fever," said a spokesman for the Moline police department. "We've handled more family fights and arguments than we normally do. Police were also called to settle domestic disputes between neighbors shoveling snow on each other's property. More than one fight broke out over the use of parking spots that had taken long, hard hours to dig out."

Others sought divine intervention. Calls to Dial-A-Prayer increased by 125 calls to 275 calls a day as people looked for comfort from the previous weekend's act of God. Reverend Fred Marsh, who managed the line, said, "I think people are down and want some help getting up, and they want to hear an encouraging voice." Reverend Marsh, who changed the messages daily to reflect the news, said that during the blizzard his prayer stated, "When it snows, we have to help one another and let others help us."

The storm predicted for Tuesday, January 16, fizzled out. Temperatures grew more moderate, and for four days the plows made headway clearing the streets around the Quad Cities. Then forecasters warned about a storm that was due to arrive January 18 and 19. Large crowds filled the Randall Foods store at Southpark Mall as shoppers stocked up for the siege. Milk and bread flew off the shelves, and there weren't enough shopping carts to go around.

While the storm itself was underwhelming, the 3.5 inches of new snow, combined with the freezing rain, created its share of headaches. Roofs collapsed, and there were a number of fender benders on icy roads. WQAD-TV lost its power for two hours. Snow and ice caused the transmitter at WOC (now my station KWQC) to lose 50 percent of its power.

After this new snow, local authorities received an estimate for what it would cost to remove the old snow. When the figures were distributed to city officials, they were asked to sit down. All told, the bill for clearing the old snow from the big blizzard in the Quad-City metro area came to a staggering $22.5 million! In other words, the 18.4 inches of snow cost $1.2 million dollars an inch. This figure reflected the expense of snow removal, any resulting physical damage, and the loss to the economy, and it was compiled with information from state employment services and state disaster relief offices in Iowa and Illinois. In the form of taxes, lost wages, or higher prices, the storm would affect the pockets of all of the citizens in the Quad Cites.

On January 23, 1979, the National Weather Service issued another winter storm warning that said 4 inches of snow "swept by northeast winds of 15 to 25 miles per hour" would cause blowing and drifting snow, which, in turn, would make traveling hazardous and difficult.

Weather-conscious members of the Quad Cities literally burned up the wires to hear the National Weather Service's recorded forecast. Jim Wiggins, chief meteorologist at the Weather Service in Moline, said the automatic answering machine that played the taped forecast became overheated from the heavy volume of calls. "We had to take the cover off it to get some cooler air circulating around it," Wiggins said. On the previous Thursday, he said that a record 2,457 people called to see if there really was another snow storm headed for the Quad Cities. Because the machine could only handle one call at a time, he said it was almost impossible to get the recording!

Another 2.4 inches of snow fell between January 23 and 24, bringing the monthly total to 25.8 inches. At this point, January 1979 was the snowiest January on record, eclipsing the record of 21.8 inches set in January 1898. The winter took third place as the snowiest in the Quad-City history, with 54.4 inches, well within striking distance of the record 69.7 inches that fell in the winter of 1975.

137

Residents carried groceries in wagons when they couldn't drive to the grocery store.
COURTESY OF *QUAD-CITY TIMES*

138

The next day, the Quad Cities were back in the deep freeze as arctic air gripped the city and the mercury dove to 10 degrees below zero. The month was now on track to become the coldest in Quad-City history. The cold also took the life of 72-year-old Hollis O. Brown, who slipped and fell on an icy sidewalk in front of his house and died of exposure.

Tired of the steady diet of snow and cold, people started looking for who or what to blame. In a letter to the *Quad-City Times,* William D. Herrstrom of New Boston, Illinois, said the Russians should be blamed. Citing Premier Nikita Krushchev's statement that the Soviet Union "would bury" the United States, Herrstrom wrote that "Soviets now have 5,000 meteorologists working on weather modification projects … Do you think you've seen too much snow? If the red devils don't pull in their horns maybe you ain't seen nuthin' yet."

Not even the experts had expected a winter of this magnitude. The summer before, Iowa climatologist Paul Waite had said the chances of having three awful winters back to back was extremely thin, and he predicted that the winter of 1979

would not be as bad as the previous two. "I was playing the odds and it certainly was off," he admitted in a subcommittee meeting of the Iowa House. I missed it by a long, long, way."

After a short break, the arctic express rolled back into town. Following half an inch of snow, the temperature slipped to 1 degree below zero on the morning of January 29. Residents took it in stride. Either they had grown to expect the worst or they were just too tired to complain about something as common as another sub-zero day.

Weariness and complacency was a good thing for those involved in clean-up operations. For weeks, Davenport residents had let it be known in no uncertain terms that the city's snow removal was unacceptable. "For weeks, we have received a blizzard of letters and an avalanche of telephone calls from irate citizens," read an editorial in the *Quad-City Times.* "We have a strong hunch that City Hall telephones have been even busier …We know we're not happy. We can't believe City Hall is pleased. It can't do much for the morale of public officials and city employees—many of whom have worked long and hard this winter—to feel the public's wrath, to be saddled with charges of incompetence and lack of concern."

In defense of his administration and city employees, Mayor Wright in the same editorial was quoted as saying, "Crews worked 24 hours a day since the New Year's Eve snowstorms and were beginning to tire. Our system was not built to handle this much snow all at once. There is no quick solution to this problem."

Out of money and short on answers, the city of Davenport had reached a breaking point. Wright traveled to Des Moines, Iowa, where he asked a committee from the Iowa House of Representatives for financial help. The state of Iowa was Wright's last resort because the Quad Cities were denied federal assistance after Governor Robert Ray had missed the federal deadline. Wright told the committee, "We find ourselves caught between Carter's administration and Ray's administration." Wright said, "I don't care who is at fault—I've got a disaster. In short, we need help."

The month ended the way it started. On the last morning of January, the temperature stood at 4 degrees below zero. A cutting west wind sent the wind chill to 30 degrees below zero.

On the record books, there had never been a month as fierce as the month of January 1979. In terms of severity and snow, that winter became the measuring stick for all other winters. The 26.7 inches of snow that fell was a January record as well as an all-time monthly record. The 18.4 inches of snow that accumulated from January 12 to January 14 was the largest amount of snow from any individual storm. A record-breaking winter total of 52.9 inches of snow fell from December through February (all but 4.5 inches fell in December and January). The snow depth of 28 inches from January 14 to 19 was the deepest ever measured in the long history of Quad-City weather.

It was a record-breaker in terms of temperature, as well. January's average temperature of 6.3 degrees was not only a monthly record, but an all-time record. The reading of 27 degrees below zero on the morning of January 2, 1979 was a daily record and the coldest temperature reading ever recorded here. Record lows were also established on January 11 (17 degrees below zero), January 14 (21 below), and January 15 (24 below). On January 2, the high of 5 degrees below zero was the coldest maximum temperature for that date. During the month of January, there were 21 days with lows of zero or below, including an 11-day stretch of below-zero temperatures.

Two decades later, that long, cold month still stands tall in the history of Quad-City weather. But this was far more than a story about blizzards and bitter cold; it was also how the people of the Quad Cities pulled together to help, comfort and nourish one another, despite the nasty comments about the plowing at city hall. Snowmobilers risked their lives to rescue hundreds of motorists stranded on windswept interstates in sub-zero cold. From Geneseo to Walcott, from Clinton to Muscatine, people opened their doors to stranded motorists, cooking meals and donating blankets or pillows. Whether it

COURTESY OF *QUAD-CITY TIMES*

was pushing a car out of a snowdrift or shoveling the walk of an elderly resident, neighbors helped neighbors. These tales of survival and hardship—especially those acts of good will performed by generous strangers—are still told today. Without this goodness and generosity, many people would not have survived the winter of 1979.

139

CHAPTER

7

The Northern Lights:
A Storm of a Different Kind

The earth's atmosphere is filled with impressive tornadoes and blizzards. But out in our solar system, there are more storms that are geomagnetic in nature, and they rage millions of miles above our heads. When they become visible, these Northern Lights are as magnificent as any phenomena known to man.

The source of any geomagnetic storm is the sun. The sun emits highly charged particles, or ions, that travel through space at speeds of up to 700 miles per second. A mass of these particles is called plasma, and as they race away from the sun, they create what is known as the solar wind. When the winds are strong enough to transport the particles of plasma into the earth's magnetic field, some collide with gases in such a way that they begin to glow, producing the spectacle that is the Northern Lights.

Known by the scientific name of *aurora borealis*, northern lights have been around since earth first formed its atmosphere. Creatures such as dinosaurs saw them, early man viewed them, and we can see them, as will our descendants.

FACING PAGE: *The Northern Lights, or aurora borealis, are a geomagnetic storm.* PHOTO © JIM REED; WWW.JIMREEDPHOTO.COM
RIGHT: *The view of the aurora as seen by the space shuttle astronauts.*
COURTESY OF NOAA/HISTORIC NWS COLLECTION

Sun Storms

Scientists have studied how the flow of particles from the sun and its magnetic field affects life on earth. Geomagnetic "storms" caused by solar activity can disrupt radio communications, endanger satellites, and knock out power systems. These sun storms are not visible from earth—except when we see the Northern Lights.

The dramatic light shows—pulsing all shades of red, blue, green, and yellow—have perplexed people throughout our history. They were chronicled by ancient Chinese astronomers, prophets in the Old Testament, and great thinkers such as Aristotle and Ben Franklin.

There is no denying the power and beauty of a sky alive with the Northern Lights. An episode gets started when energy generated by a solar flare explodes from the sun. It speeds through the solar system on the solar wind until it reaches the

Closer to the earth's magnetic poles, places like Canada and Alaska see more frequent and dramatic displays of the Northern Lights, such as these.
PHOTO © JIM REED; WWW.JIMREEDPHOTO.COM

FAR LEFT: *The time of year has no significance when it comes to your chances of seeing the Northern Lights.* COURTESY OF NOAA/CORPS COLLECTION

LEFT: *The unmatched splendor of the aurora in action.* COURTESY OF NOAA/CORPS COLLECTION

magnetic fields. In the outer reaches of our atmosphere, the highly energized particles collide with atoms of oxygen and nitrogen. The violent collisions give off the brilliant lights in every color of the rainbow.

Since the Earth's magnetic poles reside at the higher latitudes of 20 to 30 degrees, residents of Canada and Alaska have far more opportunities to witness displays of the Northern Lights than those in places like Iowa or Illinois. However, at times when exceptionally large storms erupt on the sun, the solar wind transports vast amounts of energized particles into the Earth's magnetic fields. These auroras are more vivid and powerful, which makes them visible from places farther away from the poles. It is during these times that people in the Midwest are lucky enough to see the magical show.

Predicting a display of the Northern Lights following a storm is even more difficult that predicting a tornado. The distance the energy has to travel from the sun, the time of day it arrives, the presence of clouds, and the strength of the solar wind are just a few of the factors that determine when a viewing opportunity might occur. Even so, there are various types

of predictions and forecasts that are published on the internet regarding potential times and places to view the aurora borealis. One site, www.SpaceWeather.com, provides detailed information on developing conditions or real-time sightings.

Time of year—winter as opposed to summer—has no bearing on your chances of witnessing the Northern Lights. It's all tied to the sun and its ability to kick off a solar storm—whatever time of year. The duration of an event can vary from just a few minutes to multiple hours. One thing you can do to enhance your chances of seeing a display is to set out for a spot where you have a clear view of the northern sky, free of lights, haze, and pollution.

If you've seen a good display, you know what I'm talking about. Lights arc across the sky, pulsing green, red, and yellow, bathing the sky in swoops of color. Sometimes the colors are intense and glowing. Sometimes they are muted and soft, not much more than a faint shimmer in the sky.

For my money, there is nothing that exceeds the pure beauty of the Northern Lights. The above pictures lend support to my conviction.

Weather Folklore

Northern Lights bring cold weather.
I found this to be true at least 80 percent of the time.
—TERRY SWAILS

Storm Watch: An Epilogue

Last summer, in flamboyant Midwestern fashion, a thunderstorm paid an evening visit to my corner of the world. As it huffed and puffed, I took refuge in the living room where I could get a good view of the proceedings. Staring out the window, I could barely see the old river birch across the yard as sheets of rain streamed down the glass. For half an hour, the wind swirled and the lightning flashed until the beast shuffled away, its tail between its legs.

In no time, the clouds cleared and the day ended with a breathtaking sunset, even more spectacular than the storm itself. As I watched all of this, I realized that, since the "Big Bang," similar scenes had been played out many, many times. But right now, this moment, this show belonged entirely to me.

There has always been and there will always be weather. The storms that pass through our lives are interspersed with what we call average days that pass by quiet and unmemorable. Yet, despite their unassuming nature, these days are a part of the fabric of our lives.

For many, storms are nothing more than reminders to appreciate our better days. For me, storms are drama, beauty, and grandeur. They are something to marvel at. However, when it comes to the big picture, storms are nature's way of fixing an atmosphere that is out of balance. That's one of the great things about our planet—it has the unique ability to regulate its own climate. When a monster storm revs up, it is because things are out of sync and it's too hot in one place and too cold in another. Without storms to regulate the moisture and temperatures in our atmosphere, our world would be ravaged by extremes so severe that life on earth would not be possible. For all the cancellations, delays, power outages, and tragic deaths they cause, without storms our planet would be uninhabitable. You do not have to like

storms, but you do have to respect them for what they are because they are vital to our existence.

Despite their place as atmospheric guardians, I'm more attracted to a whirling tornado or a howling blizzard because of its power and uncontrollable nature. We can land men on the moon, send pictures through phones, and even clone animals, but we can't make a storm when there's none to be had. They come when they want and go when they are ready. And, as a forecaster, I certainly know they defy prediction.

Unfortunately, storms can shatter lives and destroy the dreams and work of a lifetime. I have seen firsthand the anguish of a mother who lost a son. I've seen bulldozers clear the remains of a flooded home. I've been overwhelmed by the tears, cheers, and hugs of survivors. I've learned not to take life or nature for granted.

Because of their awesome power and destruction, I have a love–hate relationship with storms. I'm impressed by their authority but I don't want to be the beneficiary of their wrath. I want to be part of the storm experience, but at the same time, I want to survive it.

I have devoted my life to the science of weather because I want to be the first to see a storm coming. I want to live it and breathe it before it is so much as a cloud. I want to get in it and see what makes it tick. I am a weather "junkie." I'm all about the excitement a storm generates and more than that, I love being the middle guy, the one who can tell you what the skies have to offer.

So, as long as there are storm clouds on the horizon, I will be working the weather charts. I'll be searching for the next "big one," and when I find it, I'll be as passionate as ever. And when my time on earth is done, don't shed a tear for me, because the sky has always been my oyster and its storms will forever be my pearls.

Spring in the Quad Cities

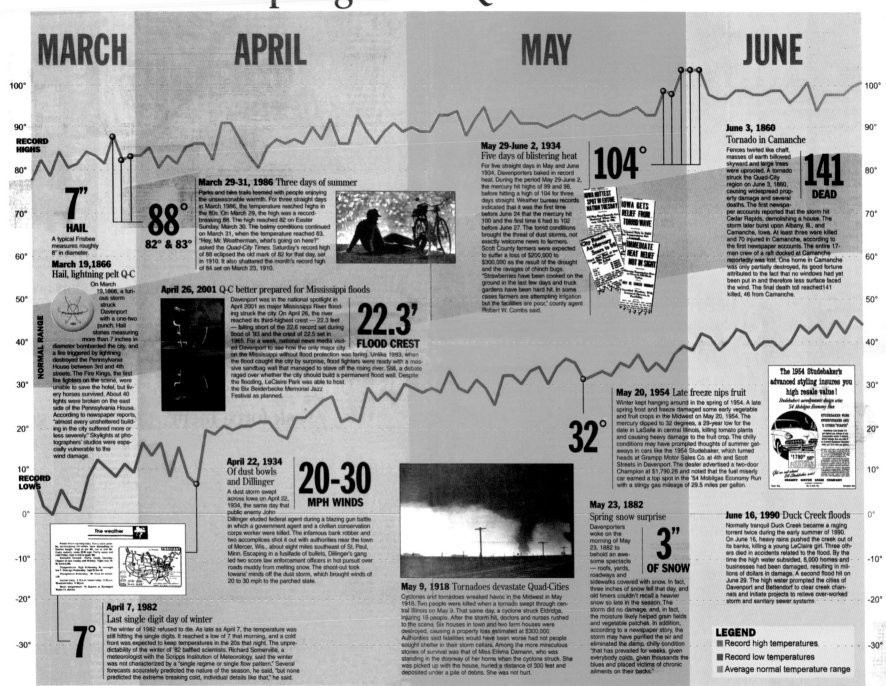

MARCH **APRIL** **MAY** **JUNE**

RECORD HIGHS — 100°, 90°, 80°, 70°, 60°, 50°, 40°, 30°, 20°, 10°, RECORD LOWS — 0°, -10°, -20°, -30°

NORMAL RANGE

7"
HAIL
A typical Frisbee measures roughly 8" in diameter.

March 19, 1866
Hail, lightning pelt Q-C
On March 19, 1866, a furious storm struck Davenport with a one-two punch. Hail stones measuring more than 7 inches in diameter bombarded the city, and a fire triggered by lightning destroyed the Pennsylvania House between 3rd and 4th streets. The Fire Kings, the first fire fighters on the scene, were unable to save the hotel, but livery horses survived. About 40 lights were broken on the east side of the Pennsylvania House. According to newspaper reports, "almost every unsheltered building in the city suffered more or less severely." Skylights at photographers' studios were especially vulnerable to the wind damage.

88°
82° & 83°

March 29-31, 1986 Three days of summer
Parks and bike trails teemed with people enjoying the unseasonable warmth. For three straight days in March 1986, the temperature reached highs in the 80s. On March 29, the high was a record-breaking 88. The high reached 82 on Easter Sunday, March 30. The balmy conditions continued on March 31, when the temperature reached 83. "Hey, Mr. Weatherman, what's going on here?" asked the *Quad-City Times*. Saturday's record high of 88 eclipsed the old mark of 82 for that day, set in 1910. It also shattered the month's record high of 84 set on March 23, 1910.

April 26, 2001 Q-C better prepared for Mississippi floods
Davenport was in the national spotlight in April 2001 as major Mississippi River flooding struck the city. On April 26, the river reached its third-highest crest — 22.3 feet — falling short of the 22.6 record set during flood of '93 and the crest of 22.5 set in 1965. For a week, national news media visited Davenport to see how the only major city on the Mississippi without flood protection was faring. Unlike 1993, when the flood caught the city by surprise, flood fighters were ready with a massive sandbag wall that managed to stave off the rising river. Still, a debate raged over whether the city should build a permanent flood wall. Despite the flooding, LeClaire Park was able to host the Bix Beiderbecke Memorial Jazz Festival as planned.

22.3'
FLOOD CREST

April 22, 1934
Of dust bowls and Dillinger
A dust storm swept across Iowa on April 22, 1934, the same day that public enemy John Dillinger eluded federal agent during a blazing gun battle in which a government agent and a civilian conservation corps worker were killed. The infamous bank robber and two accomplices shot it out with authorities near the town of Mercer, Wis., about eight miles southeast of St. Paul, Minn. Escaping in a fusillade of bullets, Dillinger's gang led two score law enforcement officers in hot pursuit over roads muddy from melting snow. The shoot-out took Iowans' minds off the dust storm, which brought winds of 20 to 30 mph to the parched state.

20-30
MPH WINDS

April 7, 1982
Last single digit day of winter
The winter of 1982 refused to die. As late as April 7, the temperature was still hitting the single digits. It reached a low of 7 that morning, and a cold front was expected to keep temperatures in the 20s that night. The unpredictability of the winter of '82 baffled scientists. Richard Somerville, a meteorologist with the Scripps Institution of Meteorology, said the winter was not characterized by a "single regime or single flow pattern." Several forecasts accurately predicted the nature of the season, he said, "but none predicted the extreme breaking cold, individual details like that," he said.

7°

May 29-June 2, 1934
Five days of blistering heat
For five straight days in May and June 1934, Davenporters baked in record heat. During the period May 29-June 2, the mercury hit highs of 99 and 98, before hitting a high of 104 for three days straight. Weather bureau records indicated that it was the first time before June 24 that the mercury hit 100 and the first time it had to 102 before June 27. The torrid conditions brought the threat of dust storms, not exactly welcome news to farmers. Scott County farmers were expected to suffer a loss of $200,000 to $300,000 as the result of the drought and the ravages of chinch bugs. "Strawberries have been cooked on the ground in the last few days and truck gardens have been hard hit. In some cases farmers are attempting irrigation but the facilities are poor," county agent Robert W. Combs said.

104°

May 20, 1954 Late freeze nips fruit
Winter kept hanging around in the spring of 1954. A late spring frost and freeze damaged some early vegetable and fruit crops in the Midwest on May 20, 1954. The mercury dipped to 32 degrees, a 29-year low for the date in LaSalle in central Illinois, killing tomato plants and causing heavy damage to the fruit crop. The chilly conditions may have prompted thoughts of summer getaways in cars like the 1954 Studebaker, which turned heads at Grampp Motor Sales Co. at 4th and Scott Streets in Davenport. The dealer advertised a two-door Champion at $1,790.28 and noted that the fuel miserly car earned a top spot in the '54 Mobilgas Economy Run with a stingy gas mileage of 29.5 miles per gallon.

32°

May 23, 1882
Spring snow surprise
Davenporters woke on the morning of May 23, 1882 to behold an awesome spectacle — roofs, yards, roadways and sidewalks covered with snow. In fact, three inches of snow fell that day, and old timers couldn't recall a heavier snow so late in the season. The storm did no damage, and, in fact, the moisture likely helped grain fields and vegetable patches. In addition, according to a newspaper story, the storm may have purified the air and eliminated the damp, chilly condition "that has prevailed for weeks, given everybody colds, given thousands the blues and placed victims of chronic ailments on their backs."

3"
OF SNOW

May 9, 1918 Tornadoes devastate Quad-Cities
Cyclones and tornadoes wreaked havoc in the Midwest in May 1918. Two people were killed when a tornado swept through central Illinois on May 9. That same day, a cyclone struck Eldridge, injuring 18 people. After the storm hit, doctors and nurses rushed to the scene. Six houses in town and two farm houses were destroyed, causing a property loss estimated at $300,000. Authorities said fatalities would have been worse had not people sought shelter in their storm cellars. Among the more miraculous stories of survival was that of Miss Emma Damann, who was standing in the doorway of her home when the cyclone struck. She was picked up with the house, hurled a distance of 300 feet and deposited under a pile of debris. She was not hurt.

June 3, 1860
Tornado in Camanche
Fences twirled like chaff, masses of earth billowed skyward and large trees were uprooted. A tornado struck the Quad-City region on June 3, 1860, causing widespread property damage and several deaths. The first newspaper accounts reported that the storm hit Cedar Rapids, demolishing a house. The storm later burst upon Albany, Ill., and Camanche, Iowa. At least three were killed and 70 injured in Camanche, according to the first newspaper accounts. The entire 17-man crew of a raft docked at Camanche reportedly was lost. One home in Camanche was only partially destroyed, its good fortune attributed to the fact that no windows had yet been put in and therefore less surface faced the wind. The final death toll reached 141 killed, 46 from Camanche.

141
DEAD

The 1954 Studebaker's advanced styling insures you high resale value!
Studebaker's aerodynamic design wins '54 Mobilgas Economy Run

June 16, 1990 Duck Creek floods
Normally tranquil Duck Creek became a raging torrent twice during the early summer of 1990. On June 16, heavy rains pushed the creek out of its banks, killing a young LeClaire girl. Three others died in accidents related to the flood. By the time the high water subsided, 8,000 homes and businesses had been damaged, resulting in millions of dollars in damage. A second flood hit on June 29. The high water prompted the cities of Davenport and Bettendorf to clear creek channels and initiate projects to relieve over-worked storm and sanitary sewer systems.

LEGEND
■ Record high temperatures
■ Record low temperatures
■ Average normal temperature range

Day 20 21 22 23 24 25 26 27 28 29 30 31 | 1 2 3 4 5 6 7 8 9 10 11 12 13 14 15 16 17 18 19 20 21 22 23 24 25 26 27 28 29 30 | 1 2 3 4 5 6 7 8 9 10 11 12 13 14 15 16 17 18 19 20 21 22 23 24 25 26 27 28 29 30 31 | 1 2 3 4 5 6 7 8 9 10 11 12 13 14 15 16 17 18 19 20 Day

Temperature Records

Day	High	Year	Low	Year	Normal High	Normal Low
MARCH						
20	77	1921	9	1965	50	30
21	82	1938	2	1960	51	31
22	78	1910	0	1888	51	31
23	84	1910	6	1960	52	32
24	80	1939	2	1974	52	32
25	81	1907	1	1960	52	32
26	84	1991	9	1955	53	33
27	83	1945	13	1964	53	33
28	81	1910	11	1970	54	33
29	88	1986	9	1887	54	34
30	82	1986	12	1969	55	34
31	83	1986	10	1969	55	34
APRIL						
1	84	2003	17	1899	56	35
2	85	1981	14	1886	56	35
3	80	1996	17	1987	56	35
4	83	1929	9	1975	57	35
5	81	1988	16	1881	57	36
6	85	1991	9	1982	58	36
7	86	1893	7	1982	58	36
8	86	1931	20	1914	59	37
9	82	1905	19	1997	59	37
10	91	1930	21	1960	60	38
11	91	1930	20	1982	60	38
12	87	1971	21	1940	60	38
13	83	1941	21	1950	61	39
14	87	2003	22	1957	61	39
15	89	2003	22	1928	62	39
16	90	2002	19	1875	62	39
17	87	2004	18	1875	62	40
18	91	2002	23	1990	63	40
19	86	1985	23	1983	63	40
20	90	1987	25	1983	64	41
21	87	1985	27	1953	64	41
22	92	1980	28	1986	64	42
23	86	1960	22	1910	65	42
24	86	1939	23	1956	65	42
25	93	1986	22	1934	66	43
26	89	1986	30	1980	66	43
27	86	1981	27	1933	66	43
28	89	1970	29	1992	67	44
29	91	1970	27	1977	67	44
30	89	1942	31	1971	67	44
MAY						
1	90	1952	30	1909	68	45
2	89	1968	28	2005	68	45
3	89	1949	25	2005	69	45
4	93	1952	28	2005	69	46
5	93	1949	34	1944	69	46
6	94	1934	30	1992	70	46
7	88	1965	29	1989	70	47
8	92	1963	32	1976	70	47
9	95	1934	30	1992	71	48
10	89	1911	26	1966	71	48
11	91	1987	33	1907	72	48
12	94	1956	33	1960	72	49
13	91	1915	31	1971	72	49
14	91	1932	34	1980	73	49
15	91	1941	35	1973	73	50
16	90	1992	31	1997	73	50
17	91	1987	35	2002	74	50
18	92	1998	33	2002	74	51
19	93	1962	36	2002	74	51
20	92	1934	32	1954	75	51
21	93	1934	32	2002	75	52
22	94	1925	34	1883	75	52
23	93	1975	33	1963	76	52
24	93	1939	36	1925	76	53
25	94	1967	35	1979	76	53
26	95	1985	38	1992	77	54
27	93	1978	34	1992	77	54
28	93	1991	36	1992	78	54
29	99	1934	34	1965	78	55
30	98	1934	38	1889	78	55
31	104	1934	40	1897	79	55
JUNE						
1	104	1934	39	1993	79	56
2	104	1934	42	2003	79	56
3	98	1972	42	1956	80	56
4	97	1934	39	1945	80	57
5	97	1933	41	1945	80	57
6	99	1933	44	1894	81	57
7	99	1933	43	1935	81	58
8	98	1963	39	1913	81	58
9	98	1911	41	1913	81	58
10	99	1933	41	1966	82	58
11	99	1933	39	1972	82	59
12	96	1987	41	1903	82	59
13	98	1987	45	1985	82	59
14	101	1987	43	1933	83	59
15	94	1894	45	1917	83	60
16	98	1918	42	1961	83	60
17	98	1897	46	1876	83	60
18	98	1994	43	1876	84	61
19	100	1953	47	1876	84	61
20	101	1988	43	2003	84	61

Summer in the Quad Cities

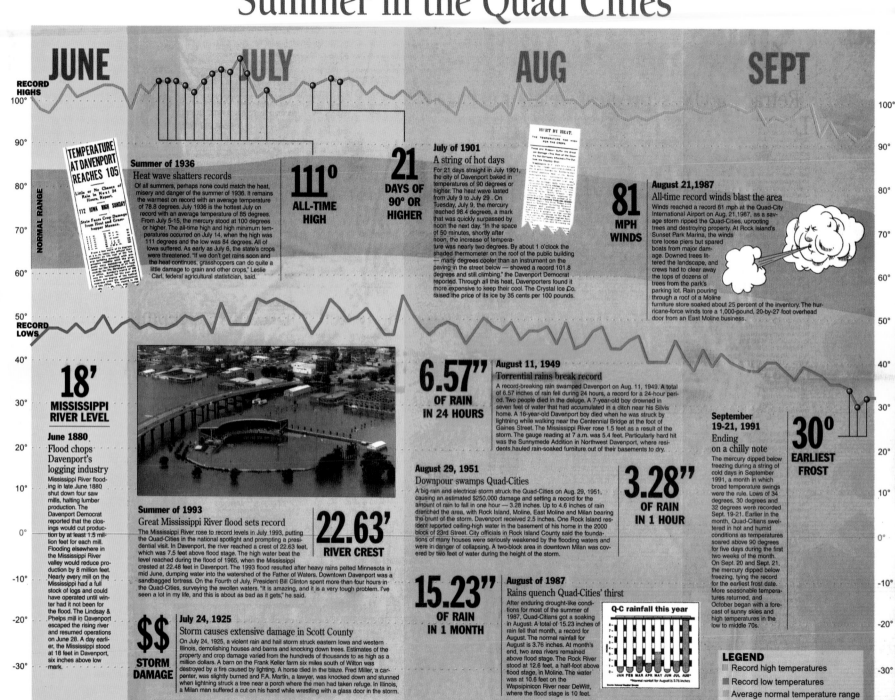

JUNE   **JULY**   **AUG**   **SEPT**

RECORD HIGHS
100°

NORMAL RANGE

RECORD LOWS

Summer of 1936
Heat wave shatters records
Of all summers, perhaps none could match the heat, misery and danger of the summer of 1936. It remains the warmest on record with an average temperature of 78.8 degrees. July 1936 is the hottest July on record with an average temperature of 85 degrees. From July 5-15, the mercury stood at 100 degrees or higher. The all-time high and high minimum temperatures occurred on July 14, when the high was 111 degrees and the low was 84 degrees. All of Iowa suffered. As early as July 6, the state's crops were threatened. "If we don't get rains soon and the heat continues, grasshoppers can do quite a little damage to grain and other crops," Leslie Carl, federal agricultural statistician, said.

TEMPERATURE AT DAVENPORT REACHES 105
Little or No Chance of Rain in Next 24 Hours, Report.
112 IOWA HIGH SUNDAY
State Faces Crop Damage from Heat and Grasshopper Menace.

111° ALL-TIME HIGH

21 DAYS OF 90° OR HIGHER

July of 1901
A string of hot days
For 21 days straight in July 1901, the city of Davenport baked in temperatures of 90 degrees or higher. The heat wave lasted from July 9 to July 29 . On Tuesday, July 9, the mercury reached 98.4 degrees, a mark that was quickly surpassed by noon the next day. "In the space of 50 minutes, shortly after noon, the increase of temperature was nearly two degrees. By about 1 o'clock the shaded thermometer on the roof of the public building — many degrees cooler than an instrument on the paving in the street below — showed a record 101.8 degrees and still climbing." the Davenport Democrat reported. Through all this heat, Davenporters found it more expensive to keep their cool. The Crystal Ice Co. raised the price of its ice by 35 cents per 100 pounds.

HURT BY HEAT.

81 MPH WINDS

August 21, 1987
All-time record winds blast the area
Winds reached a record 81 mph at the Quad-City International Airport on Aug. 21,1987, as a savage storm ripped the Quad-Cities, uprooting trees and destroying property. At Rock Island's Sunset Park Marina, the winds tore loose piers but spared boats from major damage. Downed trees littered the landscape, and crews had to clear away the tops of dozens of trees from the park's parking lot. Rain pouring through a roof of a Moline furniture store soaked about 25 percent of the inventory. The hurricane-force winds tore a 1,000-pound, 20-by-27 foot overhead door from an East Moline business.

18' MISSISSIPPI RIVER LEVEL

June 1880
Flood chops Davenport's logging industry
Mississippi River flooding in late June 1880 shut down four saw mills, halting lumber production. The Davenport Democrat reported that the closings would cut production by at least 1.5 million feet for each mill. Flooding elsewhere in the Mississippi River valley would reduce production by 8 million feet. Nearly every mill on the Mississippi had a full stock of logs and could have operated until winter had it not been for the flood. The Lindsay & Phelps mill in Davenport escaped the rising river and resumed operations on June 28. A day earlier, the Mississippi stood at 18 feet in Davenport, six inches above low mark.

6.57" OF RAIN IN 24 HOURS

August 11, 1949
Torrential rains break record
A record-breaking rain swamped Davenport on Aug. 11, 1949. A total of 6.57 inches of rain fell during 24 hours, a record for a 24-hour period. Two people died in the deluge. A 7-year-old boy drowned in seven feet of water that had accumulated in a ditch near his Silvis home. A 16-year-old Davenport boy died when he was struck by lightning while walking near the Centennial Bridge at the foot of Gaines Street. The Mississippi River rose 1.5 feet as a result of the storm. The gauge reading at 7 a.m. was 5.4 feet. Particularly hard hit was the Sunnymede Addition in Northwest Davenport, where residents hauled rain-soaked furniture out of their basements to dry.

September 19-21, 1991
Ending on a chilly note
The mercury dipped below freezing during a string of cold days in September 1991, a month in which broad temperature swings were the rule. Lows of 34 degrees, 30 degrees and 32 degrees were recorded Sept. 19-21. Earlier in the month, Quad-Citians sweltered in hot and humid conditions as temperatures soared above 90 degrees for five days during the first two weeks of the month. On Sept. 20 and Sept. 21, the mercury dipped below freezing, tying the record for the earliest frost date. More seasonable temperatures returned, and October began with a forecast of sunny skies and high temperatures in the low to middle 70s.

30° EARLIEST FROST

Summer of 1993
Great Mississippi River flood sets record
The Mississippi River rose to record levels in July 1993, putting the Quad-Cities in the national spotlight and prompting a presidential visit. In Davenport, the river reached a crest of 22.63 feet, which was 7.5 feet above flood stage. The high water beat the level reached during the flood of 1965, when the Mississippi crested at 22.48 feet. The 1993 flood resulted after heavy rains pelted Minnesota in mid June, dumping water into the watershed of the Father of Waters. Downtown Davenport was a sandbagged fortress. On the Fourth of July, President Bill Clinton spent more than four hours in the Quad-Cities, surveying the swollen waters. "It is amazing, and it is a very tough problem. I've seen a lot in my life, and this is about as bad as it gets," he said.

22.63' RIVER CREST

August 29, 1951
Downpour swamps Quad-Cities
A big rain and electrical storm struck the Quad-Cities on Aug. 29, 1951, causing an estimated $250,000 damage and setting a record for the amount of rain to fall in one hour — 3.28 inches. Up to 4.6 inches of rain drenched the area, with Rock Island, Moline, East Moline and Milan bearing the brunt of the storm. Davenport received 2.5 inches. One Rock Island resident reported ceiling-high water in the basement of his home in the 2000 block of 23rd Street. City officials in Rock Island County said the foundations of many houses were seriously weakened by the flooding waters and were in danger of collapsing. A two-block area in downtown Milan was covered by two feet of water during the height of the storm.

3.28" OF RAIN IN 1 HOUR

$$ STORM DAMAGE

July 24, 1925
Storm causes extensive damage in Scott County
On July 24, 1925, a violent rain and hail storm struck eastern Iowa and western Illinois, demolishing houses and barns and knocking down trees. Estimates of the property and crop damage varied from the hundreds of thousands to as high as a million dollars. A barn on the Frank Keller farm six miles south of Wilton was destroyed by a fire caused by lighting. A horse died in the blaze. Fred Miller, a carpenter, was slightly burned and F.A. Martin, a lawyer, was knocked down and stunned when lightning struck a tree near a porch where the men had taken refuge. In Illinois, a Milan man suffered a cut on his hand while wrestling with a glass door in the storm.

15.23" OF RAIN IN 1 MONTH

August of 1987
Rains quench Quad-Cities' thirst
After enduring drought-like conditions for most of the summer of 1987, Quad-Citians got a soaking in August. A total of 15.23 inches of rain fell that month, a record for August. The normal rainfall for August is 3.76 inches. At month's end, two area rivers remained above flood stage. The Rock River stood at 12.6 feet, a half-foot above flood stage, in Moline. The water was at 10.6 feet on the Wapsipinicon River near DeWitt, where the flood stage is 10 feet.

Q-C rainfall this year
JAN FEB MAR APR MAY JUN JUL AUG*
*Normal rainfall for August is 3.76 inches
Source National Weather Service

LEGEND
- Record high temperatures
- Record low temperatures
- Average normal temperature range

Day 21 22 23 24 25 26 27 28 29 30 31 | 1 2 3 4 5 6 7 8 9 10 11 12 13 14 15 16 17 18 19 20 21 22 23 24 25 26 27 28 29 30 31 | 1 2 3 4 5 6 7 8 9 10 11 12 13 14 15 16 17 18 19 20 21 22 23 24 25 26 27 28 29 30 31 | 1 2 3 4 5 6 7 8 9 10 11 12 13 14 15 16 17 18 19 20 21 22 Day

ALL IMAGES FROM QUAD-CITY TIMES ARCHIVES. RESEARCH BY TERRY SWAILS, STEVE GOTTSCHALK AND LYNDA BOOKER. REPORTING BY JOHN WILLARD. DESIGN AND LAYOUT BY BRAD ELLIS.

Temperature Records

Day	High	Year	Low	Year	Normal High	Normal Low
JUNE						
21	101	1988	43	1992	84	61
22	98	1911	45	1963	84	61
23	97	1931	47	1972	84	62
24	100	1937	47	1972	85	62
25	104	1988	46	1979	85	62
26	101	1931	48	1926	85	62
27	102	1934	46	1992	85	62
28	103	1934	47	0968	85	63
29	104	1936	50	1923	85	63
30	104	1931	46	1943	85	63
JULY						
1	103	1956	48	1988	86	63
2	98	1970	49	1942	86	63
3	104	1911	47	1968	86	63
4	102	1911	48	1940	86	64
5	105	1936	47	1972	86	64
6	105	1936	47	1979	86	64
7	105	1936	48	1984	86	64
8	104	1936	50	1891	86	64
9	102	1936	49	1891	86	64
10	105	1936	50	1996	86	64
11	107	1936	46	1945	86	64
12	108	1936	53	1999	86	64
13	107	1936	48	1940	86	65
14	111	1936	48	1975	86	65
15	106	1936	48	1967	86	65
16	101	1913	54	1945	86	65
17	103	1936	53	1958	86	65
18	100	1966	54	1945	87	65
19	101	1934	52	2003	87	65
20	102	1934	52	4880	87	65
21	105	1901	50	1944	86	65
22	104	1934	46	1947	86	65
23	105	1901	49	1947	86	65
24	106	1901	54	1947	86	65
25	105	1940	52	1933	86	65
26	106	1936	49	1962	86	65
27	105	1930	50	1962	86	65
28	102	1983	52	1925	86	65
29	104	1916	49	1981	86	65
30	102	1931	46	1971	86	65
31	101	1988	46	1971	86	65
AUGUST						
1	103	1987	55	1998	86	65
2	101	1987	51	1965	86	65
3	103	1930	48	1950	86	65
4	100	1947	46	1978	86	65
5	101	1918	48	1978	86	65
6	99	1947	49	1948	85	65
7	97	1984	48	1978	85	64
8	104	1913	44	1884	85	64
9	101	1934	47	1884	85	64
10	99	1983	50	1884	85	64
11	98	1941	48	1968	85	63
12	100	1936	47	1979	85	63
13	97	1918	47	2004	85	63
14	105	1936	45	2004	84	63
15	101	1936	47	1992	84	62
16	101	1988	47	1992	84	62
17	103	1988	48	1992	84	62
18	106	1936	44	1963	84	62
19	101	1983	49	1981	84	62
20	101	1983	43	1950	83	62
21	99	1936	44	1950	83	62
22	101	1936	47	1923	83	61
23	99	1947	48	1891	83	61
24	100	1936	43	1942	83	61
25	97	2003	43	1958	82	61
26	97	2003	45	1934	82	60
27	97	1976	46	1915	82	60
28	96	1953	40	1986	82	60
29	97	1984	40	1934	82	59
30	98	1953	40	1915	81	59
31	98	1953	45	1967	81	59
SEPTEMBER						
1	99	1953	41	1967	81	59
2	99	1953	38	1946	81	58
3	98	1913	43	1974	80	58
4	98	1925	39	1974	80	58
5	99	1899	42	1974	80	57
6	100	1922	40	1988	79	57
7	99	1939	39	1956	79	57
8	98	1922	38	1986	79	56
9	99	1933	38	1883	79	56
10	95	1983	40	1883	78	55
11	97	2000	40	1940	78	55
12	96	1927	38	1955	78	55
13	97	1939	36	1902	77	54
14	99	1939	39	1974	77	54
15	100	1939	38	1985	77	54
16	92	1954	35	1929	76	53
17	96	1927	35	1937	76	53
18	92	1954	65	1929	76	53
19	93	1948	34	1991	76	52
20	91	1947	30	1991	75	52
21	92	1940	32	1991	75	51
22	95	1937	32	1995	75	51

Fall in the Quad Cities

SEPT　　**OCT**　　**NOV**　　**DEC**

(Left axis, top to bottom): 100 RECORD HIGHS, 90, 80, 70, 60, 50, 40, RECORD LOWS, 30, 20, 10, 0, -10, -20, -30

(Right axis): 100, 90, 80, 70, 60, 50, 40, 30, 20, 10, 0, -10, -20, -30

NORMAL RANGE

Oct. 13, 1975
Quad-Cities basks in record October heat wave
Bikini-clad sunbathers and bare-topped yard workers were part of the Quad-City scene on Oct. 13, 1975, when a fall heat wave hit. On that day, the temperature reached 90 degrees, breaking a previous long-standing record of 85 degrees set in 1898. Record highs also were reported in numerous Iowa cities, including Sioux City, with a 92-degree reading; Mason City, 91, and Waterloo and Spencer, both with 90.

90°
HEAT WAVE

15.1"
OF SNOW

November, 1974
Snow records set in autumn month
With two snow-related records under its belt for November, the Quad-Cities ended its first December day of 1974 with up to two more inches on top of the 9.3 inches of white stuff already on the ground. As of Saturday November 30, 15.1 inches of snow had fallen during the month, about half again as much as the 10.8-inch record set in 1925. From noon Friday to noon Saturday, 8.2 inches of snow fell on the Quad-Cities, breaking the 5.7-inch record for a 24-hour period set in 1959.

Nov. 12, 1933
Dust storms swirl across the Midwest
Dust storms whipped the Midwest on Nov. 12, 1933. A worker plunged 64 feet to his death when a gust of wind blew down a scaffold at a construction site in Clinton, Iowa. Raymond L. Ross, 29, was among several killed across the Midwest when high winds and freakish dust storms pelted the region. In Davenport, a display window at the Prinz fruit market, 2nd and Ripley streets, was blown in. Two cows owned by George Ruehling were electrocuted in a field east of Brady Street and just south of the city limits when the wind wrecked a power line. The swirling dust caught the crowd at Chicago's Century of Progress, sending scores of men, women and children to the fair's hospital for eye treatment.

Nov. 29, 1974
Snow storm makes travel hazardous
An early snow storm coated Quad-City roads with ice on Nov. 29, 1974. More than 200 cars littered highways, county roads and city streets as the season's second snowstorm dumped four inches of the white stuff. Radio dispatchers in area sheriff's offices, highway patrol and police departments, harried by the strain of juggling telephones and radios, reported that at least six people were injured. Troopers warned motorists that bridges, passing lanes and isolated areas were particularly hazardous.

Dec. 5, 1910
Mississippi River drops to low levels
The Mississippi River reached its lowest stage, with a reading of 1.2 feet on Dec. 5, 1910. The weather bureau's gauge in Dubuque, Iowa, showed the river was at a stage of six-tenths of a foot below the zero reading of the gauge at 7 p.m. The zero mark was established in 1864. Ice gorges were blamed for the low readings. Forecasters said the river would rise again as soon as the water cut through the gorges.

1.2'
RIVER STAGE

-20°
& SEVERE WEATHER

Dec. 8, 1876
Blizzard postpones opening of exposition
On Dec. 8, 1876, blizzard-like conditions plunged temperatures to between 15 and 20 degrees below zero in the area. The severe weather postponed the opening of an art and industrial exposition at Turners Hall in Davenport. "Guess the ice bridge is pretty solid by this time," the Davenport Gazette reported on Dec. 9, 1876. "At 7 o'clock yesterday morning, the mercury marked 10 degrees above zero. At 6 o'clock last evening it was 6 degrees below zero."

Sept. 23, 1995
Earliest frost nips Quad-Citians
Frost on the greens delayed tee times at the Quad-City Classic for more than an hour on Sept. 23, 1995. The only frost warnings worth noting at previous golf tournaments had been those involving David Frost. Although the temperature was below par, brilliant sunshine came later in the day. The high temperature reached 59, which was lower than Quad-City Classic leader D.A. Weibring's golfing score of 65.

31°
EARLIEST AUTUMN FROST

Oct. 30, 1882
Tornado terror strikes
A hail storm and tornado struck the area on Oct. 30, 1882, bringing death and destruction. The devastation came after a stint of unseasonably balmy weather. The Davenport Democrat was filled with news of the terror. "The house of John Sharff, a 1½ story cottage, was leveled, as was the old house and his big barn . . . The kitchen of the house of J.J. Hass, a hundred rods north of Sharff's, was carried off." The total damage, the newspaper reported, "is estimated at $60,000 to $75,000, which is small amount, considering the wealth of the region through which it traveled."

Nov. 17, 1959
Cold temperatures numb Quad-Citians
The mercury dipped to a record 1 degree below zero on Nov. 17, 1959, stalling cars and jamming the Mississippi River with tons of ice. Autumn cold records that stood for nearly a century toppled in many parts of the Midwest as Arctic air fanned eastward into the Appalachians and southward into the western Gulf. A year earlier in the Quad-Cities, the temperature reached a high of 72 and dipped to a low of 44.

-1°
RECORD COLD SNAP

Nov. 12, 1934
Earthquake jolts area
A small earth quake rattled the Quad-Cities on Nov. 12, 1934. "Although houses were shaken and thousands of persons in the Quad-Cities and surrounding territory felt the two shocks no damage was reported. In Moline it was reported that the shock was so intense, however, that a street car trolley was shaken from the wire," the Davenport Daily Times reported. The newspaper was deluged with hundreds of callers who thought there had been an explosion. The tremors were attributed to a "surface quake," or a slipping of the earth's crust

Nov. 11, 1911
Cyclone pummels Davenport
A cyclone struck west of Davenport on Nov. 11, 1911, carrying barns and outhouses for distances of more than 300 feet and causing property losses of several thousands of dollars. "Run to the cellar, if you want to save your lives. There's a cyclone coming," Claus Jacobs yelled to his wife, son and daughter as the storm approached their farm one-half mile west of the Davenport mile track. He and his wife reached safety, but his daughter and son had barely reached the house when the storm overtook them. The girl was severely cut by flying glass. Five residents of Scott County suffered property damage. "The cyclone appeared in the shape of a conical black cloud and arose so suddenly that that those who lived in its center were unable to avoid its wrath," the Davenport Democrat and Leader reported.

Legend:
- ■ Record high temperatures
- ■ Record low temperatures
- ■ Average normal temperature range

(Bottom date axis): 22 23 24 25 26 27 28 29 30 | 1 2 3 4 5 6 7 8 9 10 11 12 13 14 15 16 17 18 19 20 21 22 23 24 25 26 27 28 29 30 31 | 1 2 3 4 5 6 7 8 9 10 11 12 13 14 15 16 17 18 19 20 21 22 23 24 25 26 27 28 29 30 | 1 2 3 4 5 6 7 8 9 10 11 12 13 14 15 16 17 18 19 20

ALL IMAGES FROM QUAD-CITY TIMES ARCHIVES. RESEARCH BY TERRY SWAILS, STEVE GOTTSCHALK, LYNDA BOOKER, ROY BOOKER AND THE DAVENPORT PUBLIC LIBRARY SPECIAL COLLECTIONS DEPARTMENT. REPORTING BY JOHN WILLARD. DESIGN AND LAYOUT BY BRAD ELLIS.

Temperature Records

Day	High	Year	Low	Year	Normal High	Normal Low
SEPTEMBER						
22	95	1937	32	1995	75	51
23	91	1937	31	1995	74	51
24	90	1984	30	1989	74	50
25	91	1920	30	1942	73	50
26	90	1891	32	1940	73	49
27	90	1990	30	1943	73	49
28	95	1953	24	1942	72	49
29	99	1953	30	1984	72	48
30	92	1952	28	1899	72	48
OCTOBER						
1	89	1897	30	1974	71	48
2	90	1953	24	1974	71	47
3	93	1997	27	1974	70	47
4	88	1969	27	1935	70	46
5	89	1922	28	1980	70	46
6	92	1963	24	1952	69	45
7	92	1939	27	1952	69	45
8	87	1949	25	1952	68	45
9	87	1938	26	2000	68	44
10	88	1938	24	1964	67	44
11	88	1962	24	1987	67	43
12	87	1947	26	1987	67	43
13	90	1975	23	1909	66	43
14	87	1947	26	1939	66	42
15	88	1963	21	1937	65	42
16	88	1910	26	1952	65	42
17	86	1950	23	1948	64	41
18	86	1950	19	1952	64	41
19	84	1953	20	1992	63	40
20	84	1979	22	1952	62	40
21	86	1947	20	1952	62	40

Day	High	Year	Low	Year	Normal High	Normal Low
22	84	1963	21	1976	61	39
23	85	1963	23	1981	61	39
24	83	1963	23	1981	60	39
25	81	1939	17	1887	60	38
26	85	1963	19	1962	59	38
27	84	1927	21	1976	59	37
28	84	1927	11	1925	58	37
29	84	1937	12	1925	58	37
30	84	1950	15	1925	57	36
31	85	1950	21	1913	56	36
NOVEMBER						
1	80	2000	19	1993	56	36
2	78	1938	13	1951	55	35
3	77	1938	12	1951	55	35
4	77	1978	10	1991	54	35
5	73	1945	12	1951	54	34
6	75	1916	16	1959	53	34
7	78	1915	8	1991	53	33
8	79	1999	5	1991	52	33
9	77	1999	10	1991	51	33
10	71	1949	14	1979	51	32
11	76	1964	12	1950	50	32
12	72	1902	8	1986	50	32
13	76	1989	5	1986	49	31
14	75	1971	0	1959	49	31
15	72	2001	-1	1959	48	30
16	76	1931	3	1959	48	30
17	74	1941	-1	1959	47	30
18	74	1999	7	1891	47	29
19	77	1942	5	1914	46	29
20	73	1913	8	1880	46	28
21	71	1913	1	1880	45	28

Day	High	Year	Low	Year	Normal High	Normal Low
22	69	1913	3	1937	44	28
23	65	1966	3	1950	44	27
24	66	1966	-2	1950	43	27
25	67	1896	-2	1977	43	26
26	70	1990	-6	1977	42	26
27	70	1909	1	1887	42	25
28	63	1998	-4	1976	41	25
29	69	1998	-10	1891	41	25
30	66	1998	-4	1976	41	24
DECEMBER						
1	63	1962	-8	1893	40	24
2	65	1982	-9	1886	40	23
3	69	1970	-4	1940	39	23
4	71	1998	-2	1991	39	23
5	69	2001	-2	1886	38	22
6	63	1951	-6	1882	38	22
7	66	1916	-15	1882	37	21
8	67	1991	-11	1876	37	21
9	64	1918	-17	1876	37	21
10	60	1979	-16	1972	36	20
11	67	1949	-18	1972	36	20
12	65	1991	-11	2000	35	19
13	63	1975	-15	1903	35	19
14	61	1975	-13	1901	35	19
15	57	1959	-16	1989	34	18
16	60	1939	-16	1932	34	18
17	59	1939	-15	1932	34	18
18	65	1939	-16	1985	33	17
19	62	1877	-14	1985	33	17
20	62	1877	-14	1963	33	17

Winter in the Quad Cities

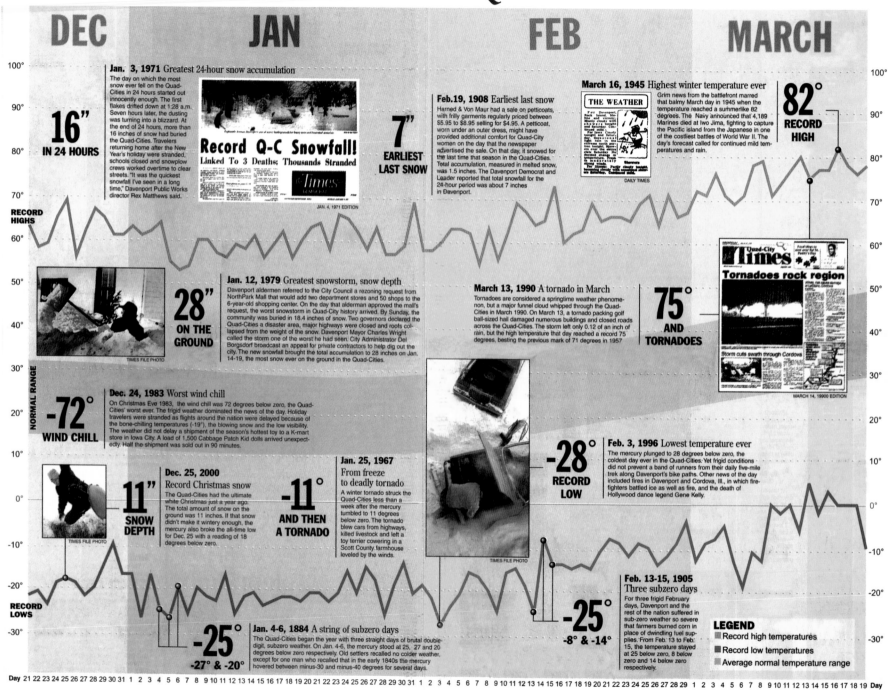

DEC **JAN** **FEB** **MARCH**

16"
IN 24 HOURS

Jan. 3, 1971 Greatest 24-hour snow accumulation

The day on which the most snow ever fell on the Quad-Cities in 24 hours started out innocently enough. The first flakes drifted down at 1:28 a.m. Seven hours later, the dusting was turning into a blizzard. At the end of 24 hours, more than 16 inches of snow had buried the Quad-Cities. Travelers returning home after the New Year's holiday were stranded, schools closed and snowplow crews worked overtime to clear streets. "It was the quickest snowfall I've seen in a long time," Davenport Public Works director Rex Matthews said.

Record Q-C Snowfall!
Linked To 3 Deaths; Thousands Stranded
Times
JAN. 4, 1971 EDITION

7"
EARLIEST LAST SNOW

Feb. 19, 1908 Earliest last snow

Harned & Von Maur had a sale on petticoats, with frilly garments regularly priced between $5.95 to $8.95 selling for $4.95. A petticoat, worn under an outer dress, might have provided additional comfort for Quad-City women on the day that the newspaper advertised the sale. On that day, it snowed for the last time that season in the Quad-Cities. Total accumulation, measured in melted snow, was 1.5 inches. The Davenport Democrat and Leader reported that total snowfall for the 24-hour period was about 7 inches in Davenport.

THE WEATHER
DAILY TIMES

March 16, 1945 Highest winter temperature ever

Grim news from the battlefront marred that balmy March day in 1945 when the temperature reached a summerlike 82 degrees. The Navy announced that 4,189 Marines died at Iwo Jima, fighting to capture the Pacific island from the Japanese in one of the costliest battles of World War II. The day's forecast called for continued mild temperatures and rain.

82°
RECORD HIGH

RECORD HIGHS

28"
ON THE GROUND

Jan. 12, 1979 Greatest snowstorm, snow depth

Davenport aldermen referred to the City Council a rezoning request from NorthPark Mall that would add two department stores and 50 shops to the 6-year-old shopping center. On the day that alderman approved the mall's request, the worst snowstorm in Quad-City history arrived. By Sunday, the community was buried in 18.4 inches of snow. Two governors declared the Quad-Cities a disaster area, major highways were closed and roofs collapsed from the weight of the snow. Davenport Mayor Charles Wright called the storm one of the worst he had seen. City Administrator Del Borgsdorf broadcast an appeal for private contractors to help dig out the city. The new snowfall brought the total accumulation to 28 inches on Jan. 14-19, the most snow ever on the ground in the Quad-Cities.

TIMES FILE PHOTO

March 13, 1990 A tornado in March

Tornadoes are considered a springtime weather phenomenon, but a major funnel cloud whipped through the Quad-Cities in March 1990. On March 13, a tornado packing golf ball-sized hail damaged numerous buildings and closed roads across the Quad-Cities. The storm left only 0.12 of an inch of rain, but the high temperature that day reached a record 75 degrees, besting the previous mark of 71 degrees in 1957.

75°
AND TORNADOES

Quad-City Times
Tornadoes rock region
Storm cuts swath through Cordova
MARCH 14, 1990 EDITION

NORMAL RANGE

-72°
WIND CHILL

Dec. 24, 1983 Worst wind chill

On Christmas Eve 1983, the wind chill was 72 degrees below zero, the Quad-Cities' worst ever. The frigid weather dominated the news of the day. Holiday travelers were stranded as flights around the nation were delayed because of the bone-chilling temperatures (-19°), the blowing snow and the low visibility. The weather did not delay a shipment of the season's hottest toy to a K-mart store in Iowa City. A load of 1,500 Cabbage Patch Kid dolls arrived unexpectedly. Half the shipment was sold out in 90 minutes.

-28°
RECORD LOW

Feb. 3, 1996 Lowest temperature ever

The mercury plunged to 28 degrees below zero, the coldest day ever in the Quad-Cities. Yet frigid conditions did not prevent a band of runners from their daily five-mile trek along Davenport's bike paths. Other news of the day included fires in Davenport and Cordova, Ill., in which firefighters battled ice as well as fire, and the death of Hollywood dance legend Gene Kelly.

TIMES FILE PHOTO

11"
SNOW DEPTH

Dec. 25, 2000 Record Christmas snow

The Quad-Cities had the ultimate white Christmas just a year ago. The total amount of snow on the ground was 11 inches. If that snow didn't make it wintery enough, the mercury also broke the all-time low for Dec. 25 with a reading of 18 degrees below zero.

-11°
AND THEN A TORNADO

Jan. 25, 1967 From freeze to deadly tornado

A winter tornado struck the Quad-Cities less than a week after the mercury tumbled to 11 degrees below zero. The tornado blew cars from highways, killed livestock and left a toy terrier cowering in a Scott County farmhouse leveled by the winds.

RECORD LOWS

-25°
-27° & -20°

Jan. 4-6, 1884 A string of subzero days

The Quad-Cities began the year with three straight days of brutal double-digit, subzero weather. On Jan. 4-6, the mercury stood at 25, 27 and 20 degrees below zero respectively. Old settlers recalled no colder weather, except for one man who recalled that in the early 1840s the mercury hovered between minus-30 and minus-40 degrees for several days.

-25°
-8° & -14°

Feb. 13-15, 1905 Three subzero days

For three frigid February days, Davenport and the rest of the nation suffered in sub-zero weather so severe that farmers burned corn in place of dwindling fuel supplies. From Feb. 13 to Feb. 15, the temperature stayed at 25 below zero, 8 below zero and 14 below zero respectively.

LEGEND
- Record high temperatures
- Record low temperatures
- Average normal temperature range

Day 21 22 23 24 25 26 27 28 29 30 31 1 2 3 4 5 6 7 8 9 10 11 12 13 14 15 16 17 18 19 20 21 22 23 24 25 26 27 28 29 30 31 1 2 3 4 5 6 7 8 9 10 11 12 13 14 15 16 17 18 19 20 21 22 23 24 25 26 27 28 29 1 2 3 4 5 6 7 8 9 10 11 12 13 14 15 16 17 18 19 Day

RESEARCH BY JOHN WILLARD, JOHN HUMENIK AND TERRY SWAILS. DESIGN AND LAYOUT BY BRAD ELLIS

Temperature Records

Day	High	Year	Low	Year	Normal High	Normal Low
DECEMBER						
21	63	1877	-22	1963	33	16
22	58	1933	-21	1963	32	16
23	59	1982	-24	1989	32	16
24	65	1889	-19	1993	32	16
25	59	1936	-18	2000	32	15
26	56	1936	-19	1962	31	15
27	62	1946	-22	1886	31	15
28	67	1984	-20	1924	31	14
29	65	1984	-15	1887	31	14
30	61	2002	-10	1909	31	14
31	61	1965	-17	1967	30	14
JANUARY						
1	63	1897	-17	1968	30	13
2	62	2004	-27	1979	30	13
3	62	1998	-17	1879	30	13
4	64	1997	-25	1884	30	13
5	55	1956	-27	1884	30	12
6	53	1933	-20	1884	30	12
7	54	1949	-26	1887	30	12
8	65	2003	-20	1875	30	12
9	60	1939	-22	1875	29	12
10	57	1975	-23	1982	29	12
11	56	1880	-17	1979	29	12
12	59	1960	-26	1974	29	12
13	58	1961	-22	1916	29	12
14	61	1928	-21	1979	29	12
15	56	1990	-24	1979	29	11
16	59	1990	-23	1888	29	12
17	62	1894	-22	1994	29	12
18	56	1996	-22	1994	29	12
19	60	1951	-23	1994	30	12
20	63	1906	-21	1985	30	12
21	64	1957	-21	1970	30	12
22	58	2002	-23	1883	30	12
23	66	1909	-21	1936	30	12
24	65	1950	-16	1884	30	12
25	62	1981	-21	1894	30	12
26	58	2002	-15	1982	30	12
27	64	2002	-18	1963	31	13
28	57	1914	-26	1963	31	13
29	57	1914	-19	1966	31	13
30	59	1988	-14	1966	31	13
31	69	1989	-21	1996	31	14
FEBRUARY						
1	58	1911	-19	1996	31	14
2	58	1987	-22	1996	32	14
3	60	1992	-28	1996	32	14
4	63	1890	-24	1996	32	15
5	63	1946	-21	1979	32	15
6	62	1882	-16	1982	32	15
7	55	1987	-17	1977	33	16
8	63	1990	-20	1895	33	16
9	56	1999	-25	1979	34	16
10	62	1976	-16	1899	34	17
11	69	1999	-15	1885	34	17
12	67	1938	-19	1899	34	17
13	64	1938	-25	1905	35	18
14	65	1954	-8	1905	35	18
15	73	1921	-14	1905	36	18
16	60	1921	-14	1958	36	19
17	62	1981	-15	1973	36	19
18	63	1913	-14	1936	37	19
19	69	1930	-15	1979	37	20
20	65	1930	-11	1929	37	20
21	66	1930	-8	1963	38	20
22	66	1922	-9	1978	38	21
23	65	1930	-12	1889	39	21
24	69	1930	-9	1993	39	21
25	71	2000	-10	1967	39	22
26	64	1971	-14	1963	40	22
27	71	1976	-12	1934	40	23
28	66	1932	-9	1962	41	23
29	67	1972	-4	1884	41	23
MARCH						
1	72	1992	-13	1962	41	23
2	71	1992	-9	1913	42	24
3	76	1983	-8	1884	42	24
4	71	1992	-9	2002	43	25
5	69	2000	-13	1960	43	25
6	73	2005	-19	1960	44	25
7	78	2000	-11	1960	44	26
8	79	2000	-13	1943	45	26
9	71	1986	0	1877	45	27
10	74	1955	-2	1934	46	27
11	77	1972	0	1948	46	27
12	80	1890	-6	1960	46	28
13	75	1990	5	1926	47	28
14	77	1995	-2	1960	47	28
15	77	1995	3	1890	48	29
16	82	1945	0	1900	48	29
17	78	1894	0	1902	49	29
18	76	1918	0	1923	49	30
19	78	1921	-10	1923	50	30

Index